Information Resources in Grant for Libraries

Information Resources in Grant for Libraries

Manoj Das Gupta

CYBER TECH PUBLICATIONS
4264/3, Ansari Road, Daryaganj, New Delhi-110002 (India)
Ph.: 011-23244078. Fax: 011-23280028
E-Mail: cyberpublicationsdelhi@yahoo.com

INFORMATION RESOURCES IN GRANT FOR LIBRARIES

Manoj Das Gupta

First Edition 2009

Published by :

G.S. Rawat for **Cyber Tech Publications**
4264/3, Ansari Road, Daryaganj, New Delhi-110002 (India)
Ph.: 011-23244078. Fax: 011-23280028
E-Mail: cyberpublicationsdelhi@yahoo.com

Printed at : Asian Offset Printers

Preface

This volume documents the grants awarded by the governments to libraries all around the world since its inception. It is being published inconnection with the thirty-sixth anniversary of the Memorial Foundation, which will becelebrated at the meeting of our Board of Trustees in Jerusalem on July 1-4, 2002.The volume provides a detailed and comprehensive picture of the MemorialFoundation's fulfillment of its mandate to reconstruct libraries, and the evolution of its programs over the last three decades.Since its inception in 1965-66, the Foundation has allocated $75,500,000 forJewish cultural and educational activities.

The government has granted $29,739,000 for11,895 scholarships and fellowships and $45,761,000 for 5,836 institutional grants. Theformer include 2,895 Fellowships, 3,034 Doctoral Scholarships, 3,234 CommunityService Scholarships and 2,733. Even more significant, these recipients represent the new generation of scholars, writers,academics, rabbis, researchers, intellectuals and artists that replaced the generation of the Indian cultural elite that were decimated in the Holocaust.

— *Author*

Preface

CONTENTS

1

Introduction

THE WORKINGS OF LIBRARY

In a parliamentary democratic system, as in India, the representatives of the people need objective, factual and timely information with a view to ensuring executive accountability to the legislature. That being the case, it is imperative that a Parliament should have its own information reservoir and information management system, away from the control of the Executive Branch. In the Indian Parliament, the multifarious information needs of members are met by Parliament's Library and Reference, Research, Documentation and Information Service, more popularly known by its acronym 'LARRDIS'.

The Parliamentary Library was established in 1921, long before India's independence, when representative government was then under the control of the administration of Britain. The Library grew in size and stature only after the achievement of national freedom in 1947. The demands of the parliamentary system contributed to the growth of the Library and in 1974 a total transformation was effected in its organizational set up when it acquired a larger role and the present name, the LARRDIS.

Ever since its inception, LARRDIS has endeavoured to live up to its role as an information provider and also as an information manufacturer. The Parliamentary Library is a key component in the LARRIDS system, with a collection of about one million volumes. The Library collections include books, debates of Indian Parliament, State Legislatures and some foreign Parliaments, reports of Central and State Governments and United Nations and its Agencies, Gazettes of Central and State Governments and other documents, including periodicals and publications brought out by the Secretariats of the two Houses of the Parliament. The collections of the Library are primarily in English. India being a nation of diverse languages, the Parliamentary Library has of course a sizeable number of books and other publications in Hindi and a large number of other national languages. Collections in foreign languages are not maintained in any significant numbers.

The traditional approach of collecting books, issuing them out to those who ask for them and answering simple queries of the customers was all that was expected of the Libraries till very recently. But the phenomenal growth in the range and dimensions of the sphere of the government has made it impossible for the modern legislator to be self-reliant in the field of information that he needs in the discharge of his duties as an effective representative of the people. The situation has necessitated the establishment of not only a well-stocked Library but also an efficient research and reference service to which Members of Parliament can always turn for help and assistance.

Remarkable developments in the field of science and technology in recent years have revolutionised all aspects of human endeavour—especially in the context of information management. People today live in an information society and without a doubt an informed electorate and Legislature are valuable assets to a successful democracy.

The advent of new technologies has not gone unnoticed or unutilized in the Parliamentary Library in India also. We have been making a modest but sustained attempt to harness the advantages of modernizing our Library so that we can serve the Members of our Parliament better.

It was in the mid-eighties that the Parliamentary Library of India initiated measures to keep pace with the technological advancements in information management. Thus, in December 1985, we made a modest beginning towards automation. A Computer Centre was set up for managing the Parliamentary Library Information System (PARLIS) with the help of the Government run National Informatics Centre (NIC). There has not been any looking back since then. In fact, within a decade and a half, we have succeeded in automating almost the entire Lok Sabha Secretariat, besides LARRDIS. The software required for different services of the Secretariat are developed in-house in collaboration with the NIC and another Government-owned firm, the Computer Maintenance Corporation (CMC) Ltd.

It may be worth mentioning that the modernization of the Lok Sabha Secretariat has indeed been a welcome development in further facilitating the varied roles of a Member of Parliament. Inevitably, the informational needs of members have also been greatly addressed with the computerisation of the Secretariat and its Branches dealing with Questions, Committees, Reporting, Members' Services, General Administration, Budget and Payment, Recruitment, Pay and Accounts, etc., and LARRDIS. Two separate but inter-linked Computer Centres are now functional-one in Parliament House and another in the Parliament House Annexe.

DATABASES

The Parliamentary Library has put special emphasis on developing its own database, since the need for objective and non-partisan information for a member is vitally important. As of now, the following are available by on-line information retrieval:

- Business of the House
    - o *Parliamentary proceedings* (Indexes of relevant parliamentary proceedings from 1985 to 1993; retrievable by the Member's name and keyword, with a facility for year-wise filtering)
    - o *Select Parliament Questions* (since 1985; lists may be generated by giving the Member's name or subject or date)
    - o *Text of the Constitution and the Constituent Assembly Debates* (List of Members and full text of proceedings)
    - o *Bills* (Index of Bills-Government & Private Members-retrievable by name of Member, subject, Bill number, amendment number, etc., available from 1985 onwards)
    - o *Practices and Procedures* [Rules of procedure and conduct of business in the Houses, Decisions/

Observations/Directions by the Chair (Indexes available, besides text of Directions from 1952; retrievable through the serial number of the Direction and subject)]

- Ministers and Members
  - o *Council of Ministers* (names and portfolios of the Union Council of Ministers since 1947)
  - o *Consolidated Index of Members* (names and membership from the time of the Constituent Assembly onwards)
  - o *Bio-data of Members* (of Ninth to Thirteenth Lok Sabha and of Rajya Sabha from 1986 onwards; statistical data relating to age, sex, education, profession, party, previous political experience and socioeconomic background of members of Lok Sabha from 1952)
- Obituary References (made in the Houses since 1921)
- Elections
  - o *Presidential and Vice-Presidential Elections* (details of all elections since 1952)
  - o *By-elections to the Lok Sabha* (since 1952)
- President's rule in States/Union Territories (since 1957)
- LARRDIS
  - o *Library Catalogue* (books/reports added since 1992; search by author, title, classification number, subject, keywords in title, keywords in context) [A programme to include all data pertaining to book acquisitions prior to 1992 is in the planning stages and we hope to have this completed soon].
  - o *Serials Control* (journals/periodicals added since 1989; title, subject, keyword searches possible); Parliamentary Documentation (this Current Awareness Service provides indexes to select books, reports and articles from 1989, retrievable under parameters like period, subject and country).
  - o *Press Clippings* (about 600 new press clippings on a wide variety of subjects available daily for on-line

scanning; search is also possible through various parameters)

- o *Library management functions* like acquisition, processing and issue and return of books have been computerized by using the software package LIBSYS. Members can access the catalogue of the Parliamentary Library through the terminals installed at the Library counters. Indexes of important articles published in newspapers and journals and publications of national and international organizations are also available through the terminals. Subject bibliographies and select lists of publications on various subjects are provided to members on request. Further expansion of the computerization efforts is envisaged to cover some other areas, including parliamentary activities like Papers tabled in the House; Pay Rolls of Members of Parliament; rare collections of Parliamentary Museum and Archives; and telecasting of parliamentary proceedings.

SUPPLY OF COMPUTERS TO MEMBERS

Keeping in view the immediate and succinct information requirements of Parliamentarians, a need was felt to provide computer facilities to them at their residences/work places. Accordingly, each Member is being provided with a computer (lap top/desk-top, as per his choice) along with all accessories like printer, data/fax modem card, etc. Free Internet and e-mail connections have also been given to each Member. A user friendly menu-driven software under the Windows environment has been developed by the NIC and the CMC which would assist the Members of Parliament in the following areas:

- Constituency functions management system: census statistics; and election statistics
- Personal information system: correspondence and grievance module; local area development system; and lists of service beneficiaries
- Office automation activities: multilingual word processing facility; e-mail facility for sending mail to fellow MPs,

Parliamentary Secretariats, Ministers and their offices, etc., fax facility, voice mail, telephone diary; greetings preparation, Directory of Who's Who; appointment planning; etc.

- Parliamentary Library Information System (PARLIS)

In order to provide on-line services such as e-mail facility and access to the parliamentary databases, other Government databases and databases of foreign Legislatures to the Members, a central computing facility has been set up in Parliament House and connected to the NIC's satellite based network, NICNET, through a Micro Earth Station (VSAT) and leased lines. Presently, Members of Parliament can have access to computerized information/databases available in the Computer Centre (in the Parliament Library) from their residences/work places through their personal computers.

Linkage

As mentioned earlier, PARLIS is connected to the NIC's satellite-based network called NICNET, which in turn is linked to the capitals of all the 32 States and Union Territories and the 600 odd District Headquarters of the country. The linkages help in the quick search for information as needed by Members. The Library is also working on a comprehensive scheme for developing a national on-line network for inter linkage of databases of PARLIS with databases of State Legislatures under a National Legislatures Information System.

Parliament of India Home Page

The Indian Parliament's Home Page, which became operational in March 1996, was accessible at the Internet site http://alfa.nic.in. The site has now been changed to http://parliamentofindia.nic.in to facilitate the search. A large volume of information is available on the Home Page. The information includes the texts of the Constitution, Constituent Assembly Debates, Select Addresses by the President of India, Rules of Procedure, Directions and Decisions from the Speaker, Bills and National Policies, bio-data of Members, current Debates of the Houses, indexes of Debates, etc. (part of the information is available from specific years only).

Microfilming

For the optimum utilization of space and better preservation of documents, the Parliamentary Library set up a Microfilming Unit in 1987. Substantial work has since been accomplished. Documents now available for computer assisted retrieval include Debates of the Legislative Council and the Central Legislative Council (representative institutions which existed during the pre-independence period), Debates in the Constituent Assembly and Lok Sabha and Rajya Sabha and the Indexes thereof, Reports of various Parliamentary Committees and the Indian Parliamentary Group, Papers tabled in the House, rare books, Government Bills, Constitution of India, Journal of Parliamentary Information, etc.

Reference Service

Within the LARRDIS is a Reference Service which is devoted exclusively to attend to the requests for information from Members. The various tools of information technology now available in the Library have helped in ensuring a speedier and better response to the queries from Members, especially on matters coming up for discussion in Parliament.

CONCLUSION

The new advances in the field of information technology, as we have seen, have enabled the expansion of the services that are made available to the Members of Parliament. Apart from this, a noteworthy yet often ignored achievement has been the speeding up of the already existing practices and procedures in the service of Members. These include the faster supply of more comprehensive information on demand, preparation of briefs and notes in anticipation of needs, etc. Application of IT has thus resulted in an overall improvement in the quality and content of the services rendered by the Library in the areas of reference and research support for Members of our Parliament.

2

INTEGRATING KNOWLEDGE MANAGEMENT TECHNOLOGIES

Technologists never evangelize without a disclaimer: "Technology is just an enabler." True enough-and the disclaimer discloses part of the problem: enabling what? One flaw in knowledge management is that it often neglects to ask what knowledge to manage and toward what end.

Knowledge management activities are all over the map: building databases, measuring intellectual capital, establishing corporate libraries, building intranets, sharing best practices, installing groupware, leading training programmes, leading cultural change, fostering collaboration, creating virtual organizations-all of these are knowledge management, and every functional and staff leader can lay claim to it. But no one claims the big question: why? (Tom Stewart in *The Case Against Knowledge Management, Business 2.0*, February 2002).

The recent summit on knowledge management (KM) at the pre-eminent ASIST conference opened on a rather upbeat note. The preface noted that KM has evolved into a mature reality from what was merely a blip on the "good idea" radar only a few years ago. Growing pervasiveness of KM in worldwide industries, organizations, and institutions marks a watershed event for what was called a fad just a few years ago. KM has become embedded in the policy, strategy, and implementation processes of worldwide corporations, governments, and institutions.

Doubling in size from 2001, the global KM market has been projected to reach US$8.8 billion during this year. Likewise, the market for KM business application capabilities such as CRM (Malhotra, 2004a) is expected to grow to $148 billion by the next year. KM is also expected to help save $31 billion in annual re-invention costs at *Fortune* 500 companies. The broader application context of KM, which includes learning, education, and training industries, offers similarly sanguine forecasts. Annual public K-12 education is estimated at $373 billion dollars in US alone, with higher education accounting for $247 billion dollars. In addition, the annual corporate and government training expenditures in the US alone are projected at over $70 billion dollars.

One can see the impact of knowledge management everywhere but in the KM technology-performance statistics (Malhotra, 2003). This seems like a contradiction of sorts given the pervasive role of information and communication technologies in most KM applications. Some industry estimates have pegged the failure rate of technology implementations for business process reengineering efforts at 70 percent. Recent industry data suggest a similar failure rate of KM related technology implementations and related applications (Darrell *et al.*, 2002). Significant failure rates persist despite tremendous improvements in sophistication of technologies and major gains in related price-performance ratios.

At the time of writing, technology executives are facing a renewed credibility crisis resulting from cost overruns and performance problems for major implementations (Anthes and Hoffman, 2003). In a recent survey by Hackett Group, 45 percent CIOs attribute these problems to technology implementations being too slow and too expensive. Interestingly, just a few months ago, some research studies had found negative correlation between tech investments and business performance (Alinean, 2002; Hoffman, 2002). Financial performance analysis of 7,500 companies relative to their IT spending and individual surveys of more than 200 companies had revealed that:

- Companies with best-performing IT investments are often most frugal IT spenders;
- Top 25 performers invested 0.8 percent of their revenues on IT in contrast to overall average of 3.7 percent; and
- Highest IT spenders typically under-performed by up to 50 percent compared with best-in-class peers.

Based upon multi-year macroeconomic analysis of hundreds of corporations, Strassmann (1997) had emphasized that it is not computers but what people do with them that matters. He had further emphasized the role of users' motivation and commitment in IT performance. Relatively recent research on implementation of enterprise level KMS (Malhotra, 1998a; Malhotra and Galletta, 1999; Malhotra and Galletta, 2003; Malhotra and Galletta, n.d. a; Malhotra and Galletta, n.d. b) has found empirical support for such socio-psychological factors in determining IT and KMS performance.

An earlier study by Forrester Research had similarly determined that the top-performing companies in terms of revenue, return on assets, and cash-flow growth spend less on IT on average than other companies. Surprisingly, some of these high performance "benchmark" companies have the lowest tech investments and are recognized laggards in adoption of leading-edge technologies. Research on best performing US companies over the last 30 years (Collins, 2001) has discovered similar "findings". The above findings may seem contrarian given persistent and long-term depiction of technology as enabler of business productivity (cf. Brynjolfsson, 1993; Brynjolfsson and Hitt, 1996; Brynjolfsson and Hitt, 1998; Kraemer, 2001).

Despite increasing sophistication of KM technologies, we are observing increasing failures of KM technology implementations (Malhotra, 2004b). The following sections discuss how such failures result from the knowledge gaps between technology inputs, knowledge processes, and business performance. Drawing upon theory, prior research, and industry case studies, we also explain why some companies that spend less on technology and are not leaders in adoption of most hyped RTE technologies succeed where others fail.

The specific focus of our analyses is on the application of KM technologies in organizational business processes for enabling real time enterprise business models. The RTE enterprise is considered the epitome of the agile adaptive and responsive enterprise capable of anticipating surprise; hence our attempt to reconcile its sense making and information processing capabilities is all the more interesting. However, our theoretical generalizations and their practical implications are relevant to IT and KM systems in most enterprises traversing through changing business environments.

DISCONNECTS BETWEEN DISRUPTIVE INFORMATION TECHNOLOGIES AND RELEVANT KNOWLEDGE

Organizations have managed knowledge for centuries. However, the popular interest in digitizing business enterprises and knowledge embedded in business processes dates back to 1993. Around this time, the *Business Week* cover story on virtual corporations (Byrne, 1993) heralded the emergence of the new model of the business enterprise. The new enterprise business model was expected to make it possible to deliver anything, anytime, and, anywhere to potential customers. It would be realized by digitally connecting distributed capabilities across organizational and geographical boundaries. Subsequently, the vision of the virtual, distributed, and digitized business enterprise became a pragmatic reality with the mainstream adoption of the internet and web.

Incidentally, the distribution and digitization of enterprise business processes was expedited by the evolution of technology architectures beyond mainframe to client-server to the internet and the web and more recently to web services. Simultaneously, the software and hardware paradigms have evolved to integrated hosted services and more recently to utility computing and on demand computing (Greenemeier, 2003a, b; Hapgood, 2003; Sawhney, 2003; Thickins, 2003) models. Organizations with legacy enterprise business applications trying to catch up with the business technology shifts have ended up with disparate islands of diverse technologies.

DECREASING UTILITY OF THE TECHNOLOGY-PUSH MODEL

Management and coordination of diverse technology architectures, data architectures, and system architectures poses obvious knowledge management challenges (Malhotra, 1996; Malhotra, 2001a; Malhotra, 2004b). Such challenges result from the need for integrating diverse technologies, computer programmes, and data sources across internal business processes.

These challenges are compounded manifold by the concurrent need for simultaneously adapting enterprise architectures to keep up with changes in the external business environment. Often such adaptation requires upgrades and changes in existing technologies or their replacement with newer technologies.

Going business enterprises often have too much (unprocessed) data and (processed) information and too many technologies. However, for most high-risk and high-return strategic decisions, timely information is often unavailable as more and more of such information is external in nature (Drucker, 1994; Malhotra, 1993; Terreberry, 1968; Emery and Trist, 1965). Also, internal information may often be hopelessly out of date with respect to evolving strategic needs. Cycles of re-structuring and downsizing often leave little time or attention to ensure that the dominant business logic is kept in tune with changing competitive and strategic needs.

As a result, most organizations of any size and scope are caught in a double whammy of sorts. They do not know what they know. In simple terms, they have incomplete knowledge of explicit and tacit data, information, and decision models available within the enterprise. Also, their very survival may sometimes hinge on obsolescing what they know.

In other words, often they may not know if the available data, information, and decision models are indeed up to speed with the radical discontinuous changes in the business environment (Arthur, 1996; Malhotra, 2000a; Nadler and Shaw, 1995). In this model, incomplete and often outdated data, information, and decision models drive the realization of the strategic execution, but with diminishing effectiveness. The model may include reactive and corrective feedback loops.

The logic for processing specific information and respective responses are all preprogrammed, pre-configured, and predetermined. The mechanistic information-processing orientation of the model generally does not encourage diverse interpretations of information or possibility of multiple responses to same information. This model of KM is often driven by technological systems that are out-of-alignment with strategic execution and may be characterized as the technology-push model.

This model has served the needs of business performance given more manageable volumes of information and lesser variety of systems within relatively certain business environment. However, with recent unprecedented growth in volumes of data and information, the continuously evolving variety of technology architectures, and the radically changing business environment, this model has outlasted its utility. The limitations of the technology-push model are evident in the following depiction of IT architectures as described in *Information Week* by LeClaire and Cooper (2000):

The infrastructure issue is affecting all businesses... E-business is forcing companies to rearchitect all or part of their IT infrastructures-and to do it quickly. For better or worse, the classic timeline of total business-process reengineering-where consultants are brought in, models are drawn up, and plans are implemented gradually over months or years-just isn't fast enough to give companies the e-commerce-ready IT infrastructures they need... Many companies can't afford to go back to the drawing board and completely rearchitect critical systems such as order fulfilment and product databases from the bottom up because they greatly depend on existing infrastructure.

More often, business-process reengineering is done reactively. Beyond its disruptive effect on business operations, most IT managers and executives don't feel there's enough time to take a holistic approach to the problem, so they attack tactical issues one-by-one. Many companies tackle a specific problem with a definitive solution rather than completely overhaul the workflow that spans from a customer query to online catalogues to order processing.

STRATEGIC EXECUTION: THE REAL DRIVER OF BUSINESS PERFORMANCE

The gap between IT and business performance has grown with the shifting focus of business technology strategists and executives. Over the past two decades, their emphasis has shifted from IT (Porter and Millar, 1985; Hammer 1990) to information (Evans and Wurster, 2002; Rayport and Sviokla, 1995; Hopper, 1990; Huber, 1993; Malhotra, 1995) to knowledge (Holsapple and

Singh, 2001; Holsapple, 2002; Koenig and Srikantaiah, 2000a; Malhotra, 2004b; Malhotra, 2000b; Malhotra, 1998c) as the lever of competitive advantage. At the time of the writing, technology sales forecasts are gloomy because of the distrust of business executives who were previously oversold on the capabilities of technologies to address real business threats and opportunities.

This follows on the heels of the on-and-off love-hate relationship of the old economy enterprises and media analysts with the new economy business models over the past decade. We first saw unwarranted wholesale adulation and subsequently wholesale decimation of technology stocks. All the while, many industry executives and most analysts have incorrectly presumed or pitched technology as the primary enabler of business performance (Collins, 2001; Schrage, 2002).

The findings from the research (Collins, 2001) on best performing companies over the last three decades are summarized in Table I. These findings are presented in terms of the inputs-processing-outcomes framework used for contrasting the technology-push model with the strategy-pull model of KM implementation.

Subsequent discussion will further explain the relative advantages of the latter in terms of strategic execution and business performance. Given latest advances in web services, the strategic framework of KM discussed here presents a viable alternative for delivering business performance as well as enterprise agility and adaptability (Strassmann, 2003).

WILL THE REAL KNOWLEDGE MANAGEMENT PLEASE STAND-UP?

The technology evangelists, criticized by Stewart (2000), have endowed the KM technologies with intrinsic and infallible capability of getting the right information to the right person at the right time. Similar critiques (cf. Malhotra, 2000a; Hildebrand, 1999) have further unraveled and explained the "myths" associated such proclamations made by the technology evangelists. Specifically, it has been underscored that in wicked business environments (Churchman, 1971; Malhotra, 1997) characterized by radical discontinuous change (Malhotra, 2000a; Malhotra,

2002b), the deterministic and reductionist logic (Odom and Starns, 2003) of the evangelists does not hold. Incidentally, most high potential business opportunities and threats are often embedded within such environments (Arthur, 1996; Malhotra, 2000c; Malhotra, 2000d). Such environments are characterized by fundamental and ongoing changes in technologies as well as the strategic composition of market forces. Increasing failures rates of KM technologies often result from their rapid obsolescence given changing business needs and technology architectures.

Popular re-labeling by vendors of many information technologies as KM technologies has not helped the situation. Skeptics of technology have observed that real knowledge is created and applied in the processes of socialization, externalization, combination, and internalization (Nonaka and Takeuchi, 1995) and outside the realm of KM technologies.

Practitioners' inability to harness relevant knowledge despite KM technologies and offices of the CKOs caused the backlash and KM was temporarily branded as a fad. Scholarly research on latest information systems and technologies, or lack thereof, has further contributed to the confusion between data management, information management, and knowledge management.

Recent reviews of theory and research on information systems and KM (Alavi and Leidner, 2001; Schultze and Leidner, 2002) seem to confirm Stewart's (2000) observation about the key flaw of knowledge management:

Knowledge Management Activities are All over the Map... But no one Claims the Big Question: Why?

Hence, it is critical that a robust distinction between technology management and knowledge management should be based on theoretical arguments that have been tested empirically in the "real world messes" (Ackoff, 1979) and the "world of re-everything" (Arthur, 1996). We are observing diminishing credibility of information technologists (Anthes and Hoffman, 2003; Hoffman, 2003; Carr, 2003).

A key reason for this is an urgent need for understanding how technologies, people, and processes together influence business performance (Murphy, 2003). Explicit focus on strategic execution as the driver of technology configurations in the strategy-pull KM framework reconciles many of the above problems.

The evolving paradigm of technology architectures to on demand plug-and-play inter-enterprise business process networks (Levitt, 2001) is expected to facilitate future realization of KM value networks. Growing popularity of the web services architecture (based upon XML, UDDI, SOAP, WSDL) is expected to support the realization of real-time deployment of business performance driven systems based upon the proposed model (Kirkpatrick, 2003; Zetie, 2003; Murphy, 2003).

The technology-push model is attributable for the inputs-and processing-driven KM implementations with emphasis on pushing data, information, and decisions. In contrast, the strategy-pull model recognizes that getting preprogrammed information to predetermined persons at the pre-specified time may not by itself ensure business performance.

Even if preprogrammed information does not become outdated, the recipient's attention and engagement with that information is at least equally important. Equally important is the reflective capability of the recipient to determine if novel interpretation of the information is necessary or if consideration of novel responses is in order given external changes in the business environment.

The technology-push model relies upon single-loop automated and unquestioned automatic and preprogrammed response to received stimulus. In contrast, the strategy-pull model has built in double-loop process that can enable a true sense-and-respond paradigm of KM. The focus of the technology-push model is on mechanistic information processing while the strategy-pull model facilitates organic sense making (Malhotra, 2001b).

The distinctive models of knowledge management have been embedded in KM implementations of most organizations since KM became fashionable. For instance, the contrast between the models can be illustrated be comparing the fundamental paradigm of KM guiding the two organizations, a US global communications company and a US global pharmaceutical firm. The telecommunications company adopted the mechanistic information-and processing-driven paradigm of KM (Stewart and Kaufman, 1995):

> *What's important is to find useful knowledge, bottle it, and pass it around.*
>
> *In contrast, given their emphasis on insights, innovation, and creativity, the pharmaceutical company adopted the organic sense-making model of KM (Dragoon, 1995, p. 52):*
>
> *There's a great big river of data out there. Rather than building dams to try and bottle it all up into discrete little entities, we just give people canoes and compasses.*

The former model enforces top-down compliance and control through delivery of institutionalized information and decision models. In contrast, the latter model encourages discovery and exploration for questioning given assumptions and surfacing new insights (Nonaka and Takeuchi, 1995).

REAL TIME STRATEGIC EXECUTION: THE REAL ENABLER OF THE RTE

The issues of technology deployment, technology utilization, and business performance need to be addressed together to ensure that technology can deliver upon the promise of business performance.

Interestingly, most implementations of KM systems motivated by the technology-push model have inadvertently treated business performance as a residual: what remains after issues of technology deployment and utilization are addressed. This perhaps explains the current malaise of IT executives and IT management in not being able to connect with business performance needs (Hoffman, 2003). A sense-and-respond KM system that can respond in real time would need to consider the holistic and collective effect of:

- real-time deployment in terms of tech and human infrastructure (inputs);
- real-time utilization in terms of what is done about or with information (processing); and
- real-time performance in terms of how it delivers business performance (outcomes).

Deployment of intranets, extranets, or, groupware cannot of itself deliver business performance. These technologies would need to be adopted and appropriated by the human users, integrated within their respective work-contexts, and effectively utilized while being driven by the performance outcomes of the enterprise.

To deliver real-time response, business performance would need to drive the information needs and technology deployment needs. This is in congruence with the knowledge management logic of the top performing companies discussed earlier. These enterprises may not have created the buzz about the latest technologies.

However, it is unquestionable that these best performing organizations harnessed organizational and inter-organizational knowledge embedded in business processes most effectively to deliver top-of-the-line results. The old model of technology deployment spanning months or often years often resulted in increasing misalignment with changing business needs. Interestingly, the proposed model turns the technology-push model on its head. The strategy-pull model illustrated in figure treats business performance not as the residual but as the prime driver of information utilization as well as IT-deployment.

The contrast between the inputs-processing-output paradigms of KM implementations is further explained in the following section to bridge the existing gaps in KM research and practice.

GAPS IN KM IMPLEMENTATION RESEARCH AND PRACTICE

The "knowledge application gap" that is characteristic of the inputs-and processing-driven technology-push model have also been the subject of criticism in scholarly research on KM (Alavi and Leidner, 2001; Zack, 2001).

However, these gaps seem to persist across most of theoretical research and industry practices related to information systems and knowledge management as shown in Table II. As discussed in Malhotra (2000a), such gaps have persisted over the past decade despite advances in understanding of KM and sophistication of technology architectures.

The sample of "definitions" of KM listed in Table II is not exhaustive but illustrative. However, it gets the point across about the missing link between KM and business performance in research and practice literatures.

Despite lack of agreement on what is KM, most such interpretations share common emphasis on the inputs-and processing-driven technology-push model. Review of most such "definitions" also leaves one begging for a response to Stewart's pointed question to technologists' evangelism about KM: "why?" In contrast, the strategy-pull model with its outcomes-driven paradigm seems to offer a more meaningful and pragmatic foundation for KM.

At least as far as real world outcomes are concerned, this paradigm measures up to the expectations about KM policy and its implementation in worldwide organizations. Better understanding of the gaps that we are trying to reconcile is possible by appreciating the contrast between the three paradigms of KM implementation that have characterized the technology-push and strategy-pull models of KM depicted in figures. This contrast is explained in terms of their primary and differential focus on the inputs, processing, and outcomes.

The inputs-driven paradigm considers information technology and KM as synonymous. The inputs-driven paradigm with its primary focuses on technologies such as digital repositories, databases, intranets, and, groupware systems has been the mainstay of many KM implementation projects. Specific choices of technologies drive the KM equation with primary emphasis on getting the right information technologies in place. However, the availability of such technologies does not ensure that they positively influence business performance.

For instance, installing a collaborative community platform may neither result in collaboration nor community (Barth, 2000; Charles, 2002; Verton, 2002). The practitioners influenced by this paradigm need to review the "lessons about technology inputs" listed earlier in Table I.

The processing-driven paradigm of KM has its focus on best practices, training and learning programmes, cultural change, collaboration, and virtual organizations. This paradigm considers KM primarily as means of processing information for various business activities. Most proponents of RTE belong to this paradigm given their credo of getting the right information to the right person at the right time. Specific focus is on the activities associated with information processing such as process redesign, workflow optimization, or automation of manual processes.

Emphasis on processes ensures that relevant technologies are adopted and possibly utilized in service of the processes. However, technology is often depicted as an easy solution to achieve some type of information processing with tenuous if any link to strategic execution needed for business performance. Implementation failures and cost-and-time overruns that characterize many large-scale technology projects are directly attributable to this paradigm (Anthes and Hoffman, 2003; Strassmann, 2003).

Often the missing link between technologies and business performance is attributable to choice of technologies intended to fix broken processes, business models, or organizational cultures. The practitioners influenced by this paradigm need to review the "lessons about processing" listed earlier in Table.

The outcomes-driven paradigm of KM has its primary focus on business performance. Key emphasis is on strategic execution for driving selection and adaptation of processes and activities, and carefully selected technologies. For instance, if collaborative community activities do not contribute to the key customer value propositions or business value propositions of the enterprise, such activities are replaced with others that are more directly relevant to business performance (Malhotra, 2002a).

If these activities are indeed relevant to business performance, then appropriate business models, processes, and culture are grown (Brooks, 1987) as a precursor to acceleration of their performance with the aid of KM technologies.

Accordingly, emphasis on business performance outcomes as the key driver ensures that relevant processes and activities, as well as, related technologies are adopted, modified, rejected, replaced, or enhanced in service of business performance. The practitioners interested in this paradigm need to review the "lessons about outcomes" listed earlier in Table I. The contrast between the outcomes-driven strategy-pull model and the input-and processing-driven technology-push model is even evident in the latest incarnation of KM under the moniker of RTE. Given the confusion between KM and KM technologies that resulted in the backlash against technology vendors, it is germane to point out a similar future for the proponents of RTE. There is an imperative need for making a clear distinction between the business performance capabilities afforded by the RTE business model and the technologies that are labelled as RTE technologies. As discussed earlier, success in strategic execution of a business process or business model may be accelerated with carefully chosen technologies. However, in absence of good business processes and business model, even the most sophisticated technologies cannot ensure corporate survival.

COMING OF THE REAL TIME ENTERPRISE: THE NEW KNOWLEDGE MANAGEMENT

The RTE enterprise is based upon the premise of getting the right information to the right people at the right time (Gartner, Inc., 2002) in "real time", *i.e.* without latency or delay (cf., Lindorff, 2002; Lindquist, 2003; Margulius, 2002; Meyer, 2002; Siegele, 2002; Stewart, 2000). Enabling the RTE should lead to faster and better decisions, and enhanced agility and adaptability. RTE represents the future of knowledge enabled business processes: wherein digitized organizations interact with increasing and relentless speed and any specific "event" results in a real-time "response".

For instance, businesses such as Gillette and Wal-Mart are trying to minimize the delay between a customer order, its shipment and the restocking of inventory with the help of radio-frequency detection (RFID) tags, also known as smart tags (Cuneo, 2003). The proponents of RTE technologies suggest that these technologies would help companies to learn to adapt, evolve, and survive within increasingly uncertain business environments.

Their rationale still seems to be based on the technology-push model of KM and may perhaps benefit from recognizing the strategy-pull model as a complement. One such perspective of RTE (Khosla and Pal, 2002) that yet does not address Stewart's

(2000) big question: "why?" and may benefit from focus proposed above is listed below:

> *Real time enterprises are organizations that enable automation of processes spanning different systems, media, and enterprise boundaries. Real time enterprises provide real time information to employees, customers, suppliers, and partners and implement processes to ensure that all information is current and consistent across all systems, minimizing batch and manual processes related to information. To achieve this, systems for a real time enterprise must be "adaptable" to change and accept "change as the process".*

The RTE will be able to operate at speeds with split-second reaction times that may far exceed human speeds of gathering and processing of information, analysis, and response (Meyer, 2002). At least, that is what the proponents of "RTE technologies" such as Khosla and Pal (2002) claim. Examples of increase of business process velocity that are often attributed to information technology include the following examples (Gartner, Inc., 2002):

- trading analytics: from 30 minutes to 5 seconds;
- airline operations: from 20 minutes to 30 seconds;
- call centre inquires: from 8 hours to 10 seconds;
- tracking finances: from 1 day to 5 minutes;
- supply chain updates: from 1 day to 15 minutes;
- phone activation: from 3 days to 1 hour;
- document transfer: from 3 days to 45 seconds;
- trade settlement: from 5 days to 1 day; and
- build-to-order PCs: from 6 weeks to 1 day.

RTE enterprises would harness everything from radio frequency sensors and smart dust to global positioning satellites and worker-monitoring software to monitor and control all processes and activities.

There are obvious benefits of the automated event-driven architectures (Sliwa, 2003) for repetitive, structured, and routine decisions (Malhotra, 2004b). Well-tested business processes may be suitable candidates for acceleration with automation of manual activities and workflows (Malhotra, 2000d). However, the more critical problem can be understood in terms of the contrast between the technology-push model and the strategy-pull model.

The programmed logic of the RTE may yield diminishing returns if environmental change outpaces the assumptions and logic embedded in its computerized networks. Split-second decisions based upon predetermined "rules" are efficient as they follow the single-loop logic and are well suited to repetitive, structured, and routine decisions.

However when such decisions are made regardless of the obsolescing business process or business model, the price is paid in terms of effectiveness (Drucker, 1994; Yuva, 2002). High-risk or high-return situations require reflection, and re-thinking as meaning of information could change and previously non-existent responses become feasible. This is all particularly applicable in contexts within which creativity and innovation facilitate emergence of new meaning, insights, and actions. Such complex meaning making and sense making capabilities for anticipating the unforeseen are yet unavailable in existing technologies (cf., Wolpert, 2001).

RTE BUSINESS MODELS: FUNCTION SHOULD DRIVE THE CHOICE OF FORM

Successful RTE enterprises focus primarily on the function of the business model that guides the choice of the infrastructure form for accelerating strategic execution. Unsuccessful RTE enterprises, in contrast, often meet their fate because of the misplaced belief that form could somehow compensate for the inadequacy of the function.

Successful RTE business models may be apparent in virtual companies such as e-Bay that owe most of their functioning to social capital embedded in their users, buyers, and sellers. Successful RTE business models may also be apparent in companies with brick-and-mortar stores such as Wal-Mart. Regardless of the variations in form, most such companies share a similar but distinctive focus on their higher purpose, which guides their strategy and its execution. This observation perhaps explains how some companies achieved most sustained business performance with lesser investments in related technologies.

Often their success was attributable to a differentiated business model based on strong ties with customers and suppliers rather than most recent investments in CRM and SCM systems. Strategic execution of the business models was accelerated with the help of technologies. However, successful companies had superior business models and a consistent track record of strategic execution as a precursor. Smart and selective investments in technologies afforded them the ability to do more with less by accelerating their business capabilities. Also, strong ties with suppliers and customers enabled them to spread the risk of investing, deploying, and utilizing the technologies with their partners and customers.

ENABLING THE RTE: ENDS SHOULD DRIVE THE CHOICE OF MEANS

The misplaced emphasis of technology-push models arose from their primary focus on the means rather than the ends as explained in this section. Most such KM implementations often happened to be caught in the convoluted complexities of technology deployment and processing without making a real difference in business performance.

Given the state of technology and the long time spans necessary for getting business systems in place, an obvious question is relevant about the superior business performers: how did the top performing companies manage to produce stellar business results despite having to choose same or similar technologies as their competitors?

It may be argued that the top performers always kept their key focus on business performance. They adopted new technologies and adapted old technologies without compromising on that primary focus. Their technologies were used for pushing data, information, and decision models just like their competitors.

However, unlike the competitors they vanquished, their choices of business processes and technologies were still driven by their primary focus on strategic execution. They may not have planned to be laggards in adopting new technologies or in spending less on such tech investments. Rather their slow but steady progress in selecting, eliminating, modifying, adapting, and integrating old and new technologies in service of their business models and business processes seemed to pay off. As they accelerated their already superior business models and business processes with new technologies, they realized greater returns in business performance. It may also be argued that many of their competitors imitated their choices of specific technologies often based upon "best practice" studies and "benchmarks" (Malhotra, 2002d).

Mistakenly treated as easy and assured solutions for fixing broken business processes and business models, new technologies further escalated the "knowledge application gap". Some of these comparison companies saw a spate of fickle and frequent technology and tech personnel changes, but their business problems persisted eventually leading to corporate failures or bankruptcies. In contrast, top performing companies have grown their business models around carefully thought out customer value propositions and business value propositions in spite of their adoption, or lack thereof, of latest technologies. Knowledge becomes the accelerator of business performance when identified with execution of business strategy rather than with the choices of tools and technologies that keep changing with time. In the eyes of the wise, knowledge and action are one (Beer, 1994).

Why do Some RTE Businesses Succeed (where others fail)?

The following cases were selected after reviewing the industry case studies of companies that were often described as benchmarks in terms of their RTE business models. Specific companies were chosen based on their visibility in the business technology press and popular media. The reviews of industry cases studies were guided by our interest in understanding the link between investments in advanced technologies and resulting business performance.

Wal-Mart: RTE Business Model where Technology Mattersless

Some IT analysts have attributed Wal-Mart's success to its investment in RTE technologies. However, Wal-Mart has emerged as a company that has set the benchmark of doing more with less. Wal-Mart did not build its competitive advantage by investing heavily or by investing in latest technologies (Schrage, 2002). A McKinsey Global Institute reports:

The technology that went into what Wal-Mart did was not brand new and not especially at the technological frontiers, but when it was combined with the firm's managerial and organizational innovations, the impact was huge.

More recently, Collins (2003) has predicted that Wal-Mart may become the first company to achieve trillion-dollar valuation within next ten years following the performance-driven model delineated in Table I and discussed earlier.

In contrast to its competitors, Wal-Mart systematically and rigorously deployed its technologies with clear focus on its core value proposition of lowest prices for mass consumers. With that singular focus, it went about setting up its supply chains and inventory management systems to accelerate business performance. Long before anyone had heard about the RTE technologies, Wal-Mart was perfecting its logistic prowess based on the hub-and-spoke model of truck routes and warehouses underlying its inventory management systems.

It was much later in the process when for its $4 billion investment in its supply chain systems its suppliers invested ten times that amount to accelerate its RTE business model underlying its supply chain network (Schrage, 2002).

The business model created the strong linkages with suppliers, which not only heavily subsidized the costs of technology investments but also pre-committed the partners to the success of the shared systems. Simultaneously, given its retail channels, distribution network, and proximity to customers through market scanner data, it has preempted its suppliers from directly competing against it.

Dell: RTE Business Model that does More with Less

Dell has developed and perfected its business model by developing strong ties with its customer base over the past 17 years. It perfected its business model over several years before accelerating its business performance with the aid of carefully selected technologies.

It has cultivated outstanding relationships with its virtual supply chain partners including outsourcing providers (such as

Solectron) and technology vendors (such as HP, Sony, and EMC). Dell also leverages its customer reach and range and market penetration for deriving commercial benefits from technologies developed by its technology partners.

It has been developing and extending the real time logic over the past several years first for selling and servicing desktop computers, and later to aggregation and distribution of value-added products and services servers, storage, networking, printers, switches, and handheld computers.

According to a survey of 7,500 companies conducted by Alinean (2002), Dell is a thrifty IT spender. Dell is equally frugal in its R&D spending (1.5 percent of revenues), according to a recent *Business Week* report, despite its continuing forays into new products and services.

Through its alliances with partners such as EMC, Dell is able to leverage their research on product innovation while itself concentrating on perfecting the linkages with customers as well as suppliers.

Dell's early innovations in passionate pursuit for being the low cost "build on demand" leader for consumer computing products has yielded it the advantage of real time business performance. More recently, it has been able to accelerate the performance of its business model with the aid of carefully chosen technologies.

GE: RTE Automation for Operational Efficiencies

GE views the real time movement as an extension of GE's renowned emphasis on Six Sigma quality drive. The business model defined for maintaining quality standards has been extended to control costs by minimizing response time to problems affecting products purchased by its customers.

GE's CIO Gary Reiner tracks once every 15 minutes what he considers to be the few most critical variables including sales, daily order rates, inventory levels, and savings from automation across the company's 13 worldwide businesses. He acknowledges that it is neither feasible nor desirable to track all kinds of information in real time even with the aid of digital dashboards.

Most operational information is tracked on daily or weekly basis while other kinds of information is tracked on an exception-reporting basis.

The company claims operational savings of 35-60 percent in costs involved in customer response, customer service, and sales. Most of these savings are attributable more to management control rather than to technologies that are used to enforce pre-negotiated contracts on its buyers who deal with its various suppliers. Operational automation that is executed in terms of command and control logic seeking compliance has not been without its adverse ramifications. GE has encountered labour management disputes resulting from the workers who are not accustomed to minute-by-minute electronic surveillance.

Cisco: Real Time Enterprise Technology Troubles

Cisco has been lauded for its RTE technologies since three years ago when its market cap was 850 percent of its recent market capitalization during this year. The company prided itself about the RTE technologies that offered apparently seamless integration of real time data within and across its supply chain and customer ordering systems. The company had legendary faith in its technologies for predictive modelling and decision-making (Carter, 2001). In a *Harvard Business Review* article, the company's CFO (of that time) claimed that: We can literally close our books within hours... the decision makers who need to achieve sales targets, manage expenses and make daily tactical operating decisions now have real-time access to detailed operating data.

Unfortunately, real-time access to data could not be of much help when, buoyed by its unparalleled growth over several quarters, Cisco made some fundamentally incorrect assumptions about the future. Cisco ignored a key lesson of KM that is often ignored by many others: the past may not be an accurate predictor of the future. While other networking companies with less sophisticated technologies had cut back on production schedules months earlier seeing impending downturn in demand, Cisco stuck to the forecasts of their "virtual close" system that they considered invincible. As Cisco (or, rather, its technology-driven forecasting systems) had never been proven wrong before, their business partners saw little merit in trying to question their proven wisdom.

As a result of misplaced faith in the power of the forecasting systems, Cisco ended up writing off $2.2 billion in inventories and sacking 8,500 employees. Industry experts and analysts suggest that Cisco's write-off resulted from its blindsided over-reliance on its much vaunted "virtual close" systems. Cisco's case demonstrates that even the best technology offers no protection against bad management decisions, especially when the assumptions embedded in the dominant logic are taken for granted. Some Cisco executives do maintain that in absence of the RTE "virtual close", the outcome could have been worse. Cisco retains its optimism in perfecting its RTE systems hoping they would eventually provide high certitude in the face of increasingly uncertain business environment.

Enron: Destroyed in Real Time

Given the dominant and intensive role of real-time information, many of the technologies associated with real-time response were initially adopted by financial services firms on the Wall Street. Given Enron Online's primary business of exchanging and trading financial data, the real-time response model seemed like a match made in heaven. Enron planned to leverage its online exchange for facilitating direct real time instantaneous transactions in the online trading of energy market commodities. In its communiqué submitted to the Federal Trade Commission, Enron had emphasized that: Efficiency gains made possible by dynamic pricing and trading are especially well suited to Enron's on-line business because electronic trading can match the speed with which commodity pricing changes. Transactions that used to take up to three minutes to complete over the phone now take just a second or two, including complex processes such as credit checks. The company deployed Tibco's vaunted "RTE platform", sought out new technology wherever possible, and planned to spend hundreds of millions of dollars on technology infrastructure.

The management control and oversight vagaries of Enron's management as well its insider-and self-dealings with fictitious entities are well documented in the records of the US Senate hearings as well as the public records of print and broadcast media. *Post-hoc* analysis of Enron's RTE technologies confirms prior observations about the technology-push model (Berinato, 2002):

> *If these [accounting irregularities] hadn't come up, the IT inefficiency might well have come up to bite Enron...*
>
> *Enron IT was as cutting edge as it was Byzantine. There were plenty of great tools, but there was precious little planning...*
>
> *The core systems supporting the main revenue-generating activities were very disjointed...*
>
> *There were major disconnects from deal capture to risk management to logistics to accounting. They all worked from different data sources...*
>
> *They had teams and teams of people who had to comb through the data and massage it so that it made sense...*
>
> *There was a lot of magic, transforming apples into oranges and oranges into apples. Preparing annual reports was a joke...*
>
> *The breakneck deployment of state-of-the-art technology was done with little regard for a management plan.*

When the cover about the collusion between Enron insiders and its auditing firm blew open, the RTE system triggered the freefall of Enron as it was also covering the risk exposure related to its instantaneous transactions.

Unfounded and overly optimistic belief in technology as the means for generating profits despite an inadequate business model led to Enron's downfall resulting in one of the largest corporate bankruptcies in US history.

CONCLUSION

This article opened with the observation that although KM activities are "all over the map" in terms of technology implementations, however, no one has asked the "big question": why? Despite diverse propositions about "getting the right information to the right person at the right time," almost everyone neglects to ask what knowledge to manage and toward what end. A review of the industry case studies of companies characterized in the recent years as RTE business enterprises surfaced some interesting insights. Recent industry analyses that have demonstrated inverse correlations between IT investments and business performance provided some motivation for this analyses.

Additional impetus was provided by the contrast between the hype about "RTE technologies" propagated by some IT analysts and our in-depth analysis of companies that achieved success as RTE benchmarks. To some extent the search for the "next big thing" and the "killer app" is to blame for its narrow focus on IT and innovation as ends rather than means for achieving sustainable business performance (*Business Week,* 2003). The big question "Why?" should drive tactical and operational aspects of technology and process related innovations in an organizational KM implementation. As contrasted with the inputs-and processing-focused technology-push model, explicit and specific performance outcomes oriented focus of the strategy-push model, further emphasized the focus on the "big question."

The contrast between the three archetypes of inputs-, processing-, and outputs-driven paradigms of KM explained in Table I and Table II further aided deconstruction of the existing conceptualizations and practices of KM. One such conceptualization of KM that has been applied in diverse worldwide governmental and corporate practices was then discussed to motivate subsequent discussion on the RTE business models.

The contrast between information-processing capabilities of latest technologies and needed sense-making capabilities was then explained. Additionally, the mechanistic emphasis of technology-based linkages was contrasted with appreciation for organic and socio-psychological relationships needed for nurturing knowledge processes. Two propositions were offered based on prior discussion-one pertaining to the form and function of the RTE, and the second relevant to the contrast between ends and means of achieving performance outcomes. Based upon original analyses, review of prior research, and review of industry case studies we made specific managerial recommendations about realizing the real time performance of enterprise business models.

Specifically, we recommended that:

- organizational function should drive the choice of organizational form; and
- ends should drive the choice of means.

The above propositions were then illustrated with the aid of RTE industry case studies that have been used by IT analysts to hype the benefits of RTE technologies. Based upon our analyses, we counter-argued that the benefits attributed to the RTE technologies should indeed be attributable to the RTE business model. We further contended that in absence of an effective RTE business model, even the most expensive and sophisticated technology could not ensure corporate survival in the short-or long-term. The RTE case studies lent support to the primary role of strategic execution as the lever for sustained business performance.

As discussed, the successful RTE enterprises achieved their success by staying a step ahead of competition and by offering value propositions that inspired customers' imagination instead of playing the "me too" game in an already crowded market. These companies also selected and integrated ICT capabilities that fit directly with what they were deeply passionate about, what they believed they could be the best at, and what directly drove their steady economic growth. The successful RTE businesses did not adopt new technologies motivated by fear of getting behind. Rather, they thought differently about technology as an accelerator of business momentum and not its creator.

Unlike the successful models of RTE enterprises, the failures were characterized by thoughtless reliance on technology often grasped as an easy solution, without coherent understanding of how it links to strategic execution for business performance.

3

Non-print Materials and Other Formats

Microforms: Microforms of collections of printed works, produced by commercial publishers, are considered for purchase when this is the only realistic way of acquiring the material. However, electronic resources are preferred. Microform collections are not normally purchased if the original material is readily accessible in a major library in the UK.

Microforms of manuscript collections, produced by commercial publishers, are considered for purchase but are not normally acquired if the originals are readily accessible in a major library or archive in the UK.

Microforms of individual monographs are rarely purchased but exceptions may occasionally be made for works of particular importance when the hard copy is unavailable.

Audio and Audio-visual Materials: The University Library does not actively collect material issued in any of these formats, but cassettes and computer disks are often received in conjunction with a printed item received by legal deposit or by purchase.

Audiobooks, although not liable to legal deposit, are nevertheless often received independently of the printed work.

ELECTRONIC PUBLICATIONS

The University Library's electronic collection aims to support research and teaching in the university in all subject areas by providing access to publications, bibliographic and full-text, where the electronic version offers advantages over print, for example by widening access or providing improved search and retrieval or analysis capabilities. Although the University Library's collecting policy is broadly format-neutral, decisions on the acquisition of individual titles take into account long-term preservation and access factors affecting the format.

The University Library will not normally acquire both the electronic and print versions of publications where the content is substantially the same, except where the print version is already in the collection and access to the electronic can be justified on the basis of improved functionality.

The University Library may seek to acquire access to electronic publications in conjunction with one or more departmental or faculty libraries, where appropriate to their subject interests.

Full-text electronic publications acquired include: electronic journals, reference works (especially encyclopaedias and dictionaries), full-text literature databases, historical documents, and maps. Electronic versions of rare books and manuscripts are acquired to facilitate consultation of fragile material, support preservation of the originals, and provide access to important collections available in other libraries and archives. Electronic full-text may also be acquired to support cross-searching of books, manuscripts and documents.

Electronic versions are acquired of national broadsheet newspapers, current and retrospective and their indexes.

Electronic books are acquired to support research unless received under legal deposit. It is expected that undergraduate level electronic books will be acquired by the faculties, departments and colleges.

A substantial proportion of official publications issued by governmental and non-governmental organisations previously acquired in print formats are freely available online, either to the general public or through national data centres such as the Economic and Social Sciences Data Service to UK higher education. The main library acquires further electronic publications in this area, notably those issued by NGOs (including the OECD and United Nations) through subscription where relevant to research interests in the university.

In addition to subscriptions to electronic publications within the JISC collection the University Library is entitled to access to a number of titles in the JISC collection made freely available to UK higher education. This includes research and teaching material, both text and multimedia, held by national data centres such as MIMAS and EDINA.

Rare Books (Including 'Cam')

The Rare Books department purchases books and pamphlets in European languages printed before 1850. This range covers periodical publications, often in collected editions, but including, for example, individual early newsbooks and runs of periodical issues. Some purchases contain music (play-texts, opera librettos) and maps (topographical and travel books) but the chief responsibility for antiquarian music rests with the Music department and for maps and atlases with the Map department. While all European languages, ancient and modern, are within the remit of the Rare Books department, the main modern languages predominate.

The general principle for book selection is that of building on the existing strengths of the collections, thus increasing their already rich research potential. Purchases are concentrated on works of substantial interest. The range of these collections is very wide, but the following categories are of special importance:

1) incunables and post-incunables, *i.e.* books printed up to 1520 (These tend to be expensive, and hence occasional, purchases.)
2) works of significance for classical scholarship and its history
3) architectural books
4) the history of science, mathematics, medicine and psychiatry
5) books of special relevance to Cambridge, *e.g.* the history of the university, city and county, and important Cambridge figures
6) books about Ireland, and books printed in Ireland
7) English literature and political history of the Restoration and early eighteenth century

8) French and German literature and philosophy of the eighteenth and early nineteenth centuries
9) the history of the book and the book-trade, typography and book-illustration
10) special author collections, especially Swift, Sterne, Rousseau and Goethe

Care is taken to avoid purchasing books of which there is a copy in another Cambridge library. Books are not purchased for the sake of their provenance or binding, but from time to time copies are purchased because they contain significant annotations. Otherwise duplicate or near-duplicate copies of an edition are not purchased, nor are multiple editions of a work, unless they are known to contain substantial textual variations.

Translations of foreign works into English are often purchased; in other instances only translations of major works or translations by major figures are purchased.

Books are not normally purchased if the main library already holds a microfilm or facsimile.

With few exceptions books which are in poor condition or which appear to be overpriced are not purchased.

In exceptional circumstances the negative factors referred to in 27.3-27.6 may be overturned for items falling within the categories of special interest in 27.2.

The main library subscribes to the digital resource Early English Books Online, which contains versions of virtually all English books printed from 1473 to 1700. There is also a subscription to the Eighteenth Century microfilm series and its digital version Eighteenth Century Collections Online, both of which cover English books of that period, and to Incunabula: The Printing Revolution in Europe 1450-1500, a selective microfiche series. Coverage of continental books on microfilm is very patchy and this fact together with their fairly easy availability and rather lower pricing has directed the acquisition of originals in the past decade more strongly toward continental books.

A number of areas of special research value derive from, or have been enhanced by, the acquisition within the last twenty years of particular collections-for example, the history of psychiatry (Hunter), English Restoration literature and politics (Brett-Smith, Verney). The financial resources for this kind of collection development may well be less easily found in the future and so only collections with a duplicate ratio of less than a third are seriously considered for purchase. (A somewhat higher ratio may be allowable for deposited or donated collections.) Nevertheless it has sometimes been possible to establish the main library as a pre-eminent centre for research in a particular subject by the acquisition of a single collection, and this possibility forms part of the policy for the future.

'Cam' and 'Cam Papers'

The main library aims to acquire a comprehensive collection of books and pamphlets relating to the university and city of Cambridge and to the historic county of Cambridgeshire (*i.e.* including the Isle of Ely, but excluding Huntingdonshire and the Soke of Peterborough). This collection (class 'Cam') also includes material relating to the new larger Cambridgeshire as a whole and to the wider geographical area in which Cambridgeshire predominates, which might be described as the Fens or 'mid-Anglia'. Material relating specifically to places included within the boundaries of the larger Cambridgeshire but outside the boundaries of historic Cambridgeshire (*e.g.* Huntingdon, St Neots, Peterborough) or to places immediately adjacent to the county (*e.g.* Royston, Newmarket) is also acquired but is not added to class 'Cam'.

Class ' Cam' includes material whose content relates to the geographical area defined in 27.10.1, including works of fiction set in the county, whenever the setting is significant rather than merely incidental. Material suitable for class ' Cam' is almost always claimable by legal deposit and much of the collection is acquired in this way. Some material is purchased for the collection, for example if legal deposit claims have failed or if an item is considered to be of such restricted local interest that the other legal deposit libraries would not wish to receive it. Such purchases are almost always inexpensive. In the case of academic works of which the main library needs a freely borrowable copy a second copy is purchased for this purpose. 'Cam Papers' includes a range of ephemeral material, generally outside the scope of legal deposit, in which the main library is interested solely because of its local nature. The collection includes posters, notices, leaflets and programmes issued by university or college societies, or in conjunction with various official or unofficial lectures, seminars, courses and meetings, theatrical or musical productions, memorial services etc. The focus of the collection is the university and colleges rather than the city and county. Material in this collection is acquired by donation. Ephemera of an official nature forms part of the university archives. The main library actively acquires material at the Societies' Fair, held at the beginning of each academic year. Other material is acquired more haphazardly, whenever members of staff come across individual items, or when officers of university societies donate material. The visits of library staff to the Societies' Fair are also used to publicise the existence of the collection and to encourage societies to deposit or donate their papers. The minute books and correspondence of university societies, and their other official records, are acquired whenever possible. These are handled by the Manuscripts department.

Portraits and Photographs: The main library has a collection of portraits (prints, photographs and reproductions of paintings; class 'Portraits'). A large proportion has Cambridge connections, and also a few paintings. In addition to this the E.H.L. Jennings Collection contains c. 100,000 portraits of historical personages. There is no general collection of photographs, although a number of archives include photographs, often relating specifically to Cambridge. Other collections, such as the Vickers archive and, most notably, the Royal Commonwealth Society collection, contain photographic materials. The main library does not normally seek to acquire portrait or photographic materials, except when they complement an existing collection (*e.g.* Royal Commonwealth Society), but such materials are received from time to time as part of larger collections.

MANUSCRIPTS AND ARCHIVES

The main library contains very extensive collections of manuscripts and archives which have been accumulating since the thirteenth century. High priority is attached to adding manuscripts and archives of research interest to the collections. Oriental manuscripts are acquired selectively in accordance with the same criteria as western manuscripts.

The main library seeks to acquire only original and unique material and would not, for instance, acquire a collection of photocopies of papers housed elsewhere. Nor would the main library acquire collections of printed ephemera, press cuttings and the like, although occasionally material of this type is accepted as part of larger collections. Collections deposited on loan are rarely accepted, although exceptions are occasionally made, especially for material transferred from elsewhere within the university.

The manuscripts and archives which are acquired normally fall into one or both of the following categories:

1) material concerning the university of Cambridge and its members
2) material which enhances or complements the existing manuscript holdings

The first category includes, at one extreme, the university's official records (which go into the university archives) and, at the other, private papers of individuals who studied at Cambridge. Papers concerning the city and county of Cambridge are not normally acquired as these are more properly the concern of the County Record Office.

The second category is a large one because of the range and variety of the main Library's holdings. Examples of acquisitions in this category would be additional papers of a person or corporate body whose papers are already held, or a manuscript in a field in which the main library is already strong (*e.g.* Middle English manuscripts).

The assessment of potential acquisitions also takes into account the amount of space they will take up and what resources in terms of making the material accessible to readers, cataloguing, conservation and administration they are likely to absorb. Where material is declined a more appropriate home may be suggested. The application of the principles outlined above varies somewhat according to whether material is acquired by purchase, donation or loan. For example single autograph letters are not normally purchased but would be accepted as a donation.

Theses approved for the award of Cambridge higher degrees are deposited in the Library. These include all Ph.D., M.Litt., M.Sc., M.D. and M.Chir. theses, as well as those submitted for other higher degrees (*e.g.* D.D.). No undergraduate dissertations are required to be deposited, but those submitted for the B.A. degree in geography, and which are awarded first-class honours, are transferred to the main library from the Department of Geography, normally 3 years after submission. Some prize essays and lectures in the university are required by Ordinance to be deposited in the main library. Other prize essays and lectures, although not subject to deposit regulations, are also received by arrangement, albeit selectively.

DEPENDENT LIBRARIES

Science Libraries: The University Library's current working collections in the sciences (excluding clinical medicine and clinical veterinary medicine) are located in the Betty and Gordon Moore Library (hereafter Moore) and the Central Science Library (hereafter CSL). The long-term retention and preservation of stock in these subject areas is the responsibility of the main library. The primary function of the science libraries is to provide the information needed to support the research and teaching needs of the university in the subject areas outlined below:

- *Moore:* pure and applied mathematics (including mathematical statistics), physics, astronomy, computer science, materials science, and engineering.
- *CSL:* chemistry and chemical engineering, earth sciences, hydrology, remote sensing, environmental sciences, biological sciences (plant sciences, zoology, biochemistry), preclinical biomedical sciences (pathology, physiology, pharmacology, genetics, anatomy), neuroscience, and microscopy (with the exception of purely technical material or applications to engineering or materials science, which is collected by the Moore).

Additionally, the Moore is the main library for the Faculty of Mathematics, while the CSL holds the historical collections of the

Cambridge Philosophical Society. The science libraries acquire material (predominantly journals and monographs) chiefly at advanced academic, research, and professional levels. Such material received by the main library through legal deposit in appropriate subject areas is selectively located in the science libraries, and this intake is complemented by purchased material. A significant number of journal titles is acquired through the CSL's exchange programme, while donations and transfers may be accepted if they contribute to the science libraries' ability to satisfy user needs.

Historical, sociological, biographical, and elementary/popular material in the sciences, as well as interdisciplinary material falling within the remit both of the science libraries and the main library, is retained by the main library. The most appropriate location for interdisciplinary material whose subject matter would qualify it for selection by two or more dependent libraries is determined via discussion amongst the librarians involved.

The science libraries are in principle single copy libraries, but additional copies may be acquired where a particular text is heavily used or heavy use is anticipated. In such instances, the location of high-demand material in either the short loan or reference classes is preferred to the purchase of large numbers of multiple copies. The majority of new acquisitions are in English, this being the standard language of scientific communication, but material in major European languages is also acquired.

The science libraries' policy of open access to their monographic collections means that all paperback monographs acquired through legal deposit which are to be located in either library must be bound on arrival.

MEDICAL LIBRARY

The Medical Library is the central library for staff and students working on the Addenbrooke's site, supporting the education, research and health-care practice activities of the university (especially the Faculty of Clinical Medicine and the Addenbrooke's-based departments of the Faculty of Biology B), Medical Research Council units, and the Addenbrooke's and other local NHS Trusts. It also serves NHS staff in the community and the region. It does not support basic nurse education, but does support post-basic courses for qualified nurses and midwives. It does not provide information for patients or the public. It is primarily a working library, with an emphasis on the currency of its collections and resources.

The Medical Library interprets 'collection development' to embrace printed materials and other media. The latter cover, to an increasingly important degree, electronic resources served both from within the library and from external sites, including but not restricted to multimedia, CD-ROMs, and remote databases. Audio-visual formats (videocassettes, audiocassettes and slides) are no longer acquired.

Journals constitute the most important part of the medical literature, and account for the major part of the Medical Library's collections. Titles are selected at advanced academic, research and professional levels. Subscriptions are usually approved only when the title in question has been recommended by a user of the Medical Library. Legal deposit titles are located in the Medical Library when staff in the main library and Medical Library staff judge this to be appropriate.

External resources are in some cases governed by contractual restrictions limiting access to specific user communities such as higher education or the NHS. In these cases the same resources may be duplicated across more than one contract to ensure that different user communities have access to broadly similar resources.

The Medical Library acquires monographic and reference material at advanced academic, research and professional levels. Teaching texts, including multiple copies of standard works, MCQs, and test papers are acquired for clinical and other health sciences students. Books received on legal deposit in the main library are selectively located in the Medical Library to complement the latter's collections. Popular-level works are not normally acquired.

The collection is classified according to the US National Library of Medicine scheme (supplemented by Library of Congress) and is focused on clinical medicine. Titles are purchased covering virtually the entire spectrum of health-care practice, *viz.* medicine, surgery, public health, psychiatry, diagnostic imaging, medical laboratory sciences, nursing, allied health sciences, and hospital administration. Other areas, such as dentistry, are covered only by the acquisition of legal deposit material.

Preclinical medical education is not supported, being the responsibility of the School of Biological Sciences' departmental libraries in central Cambridge. However preclinical sciences-anatomy, physiology, and especially biochemistry, microbiology, pathology, and pharmacology-are an essential element in medical laboratory sciences and so are covered as a necessary adjunct to clinical practice.

The history of medicine section includes secondary texts on the subject, together with primary material relating to Cambridge and its contribution to medicine. This collection is intended to supplement those at the main library and the Whipple Library.

In addition to clinical texts, there is a small collection of general classical and contemporary non-medical literature to support the study of medicine and the humanities. Virtually all acquisitions are in English, this being the standard language of scientific communication. Donations and transfers are accepted in accordance with the University Library's overall policy.

In keeping with the Medical Library's primary role as a current working library, its stock is regularly weeded to ensure the currency of its core collections. Stock thus identified, if received by legal deposit or a unique holding in Cambridge, will normally be relocated to other parts of the University Library; other duplicate material will be discarded.

SQUIRE LAW LIBRARY

The Squire Law Library is one of the largest academic legal libraries in the United Kingdom and houses the primary law collection for the University of Cambridge. The aims of the Squire Law Library, and with reference to its collection building, are four fold:

- to support the teaching and research aims of the Cambridge Faculty of Law and to help sustain the Faculty's national and international reputation as a centre of excellence in legal studies.

- to respond to the needs of new teaching courses and evolving research interests of the Faculty and its various Centres. In broad terms the library aims to respond in its collection building to the changing law and the many evolving legal systems around the world.
- to contribute to the mission of Cambridge University Library as one of its four dependent libraries.
- to contribute to the building of the University Library's world-class research collection by maintaining a top quality academic law library that incorporates national, international and historical material.
- to act as an element in the legal deposit system where relevant law-related material is housed at the Squire.

The Squire's collections support the undergraduate Tripos and postgraduate LL.M programmes by ensuring that sufficient copies of key texts and other publications are available, where resources permit, in order to allow students to have adequate access to the material provided on their reading lists. The library also aims to maintain and develop collections of a broad nature that support and reflect the research interests of the law academics, the PhD law students and any official visitors to the Faculty of Law.

The Squire Law Library aims to develop its collections by embracing relevant printed materials and electronic versions (through subscription services – *e.g.*. Lexis, WestLaw UK, HeinOnline-and free internet gateways – *e.g.*, BAILII, Intute) of legal material covering primary and secondary legal sources of law. CD-ROMs, microfiche and microfilm are also acquired where appropriate.

Materials that are considered in the acquisitions selection process include reference works, parliamentary papers, government reports, treaties, legislation, abridgements and commentaries, journals and law reviews, loose-leaf works, textbooks and monographs. The collecting of selected antiquarian law books and reports is also included as part of the acquisitions policy. The library collects mainly academic teaching and research books and publications that are considered appropriate for undergraduate or advanced research work. Practitioner works are obtained only where requested for teaching and research purposes. "A" level texts relating to the study of law are not included.

The Squire maintains a full academic collection relating to the law of England and Wales, Scotland, Northern Ireland and the Republic of Ireland. The Squire collects extensively in the countries with a common law tradition including the USA and the Commonwealth Countries (including the major jurisdictions of Australia, Canada, India, New Zealand and South Africa).

For comparative purposes the library also aims to build research collections for all civil law jurisdictions of Europe. The library has always maintained in depth sections for France, Germany and Italy but also aims to keep current primary and secondary materials for other European countries.

The collection development policy also extends to the following headings: Public International Law, European Union Law, Roman Law, Canon Law, Legal History, Jurisprudence, Medical Ethics, Conflict of Laws and Comparative materials that are concerned with more than one jurisdiction.

Within each jurisdiction and collection heading and, where financial resources permit, the library aims to acquire materials relating to all aspects of the law (both public and private law) that are of interest to the Faculty and its student and research population.

The majority of works are maintained in the English language, though texts in French, German and Italian are also acquired both where their jurisdictions are concerned and also in relation to the European Union, Conflict of Law and Public International Law collections. Other materials are acquired in the vernacular when recommended or where received as donations.

The method of acquisition falls into four main categories, (i) purchases using major agents, suppliers and publishers (ii) legal deposit via the main library where books and serials are concerned (iii) donations received from individuals, organizations and, where review copies are concerned, the editorial committee of the Cambridge Law Journal (iv) Cambridge Law Journal Exchange programme for law reviews by other Law Schools world wide.

Materials are selected for acquisition via recommendation from Faculty members, Tripos and LL.M course reading lists, interest from researchers (including law PhD students), publishers' catalogues, legal literature, library literature and bibliographies, accessions lists and current awareness services. Items for acquisitions are evaluated according to quality, relevance, interest, cost, availability, language and format. Consideration is given to the number of copies needed or the relevant methods of access in order to allow appropriate provision to the user.

The Squire Law Library's collection development policy takes into account the interests of other Cambridge libraries (*e.g.* Criminology, Economics, History and Management Studies) and coordination regarding acquisitions that overlap across related disciplines does occur where it is feasible to do so. Some duplication is necessary where specific teaching and research occurs in more than one Faculty (*e.g.*. in subject areas like corporate governance or tax law).

In order to ensure a more complete coverage given the limitation of resources, coordination at a national level is regularly discussed and explored with other leading law libraries such as the Institute of Advanced Legal Studies (London) and the Bodleian Law Library (Oxford).

SECTION ON LIBRARY AND RESEARCH SERVICES FOR PARLIAMENTS

Information Management in the Indian Parliament

The Origins: The Parliament Library and Reference Research Documentation and Information Service (LARRDIS) has been established to meet the information requirements of the Members of both Houses of Parliament in India. Though the nomenclature (and the array of activities it signifies) is of recent vintage, the Parliament Library has a long history, beginning from 1921, during the days of the Central Legislative Assembly, when India was still under British rule. During the pre-independence days the Library remained a very small establishment meeting the limited requirements of the Members of the Central Legislative Assembly.

It was only after Independence in the year 1947 and after the Constituent Assembly commenced the work of drafting a new

Constitution for free India, that the demands on the Library started multiplying. Too much dependence by Members only on governmental sources of information for policy-making as well as evaluation of policy, or even for critical appraisal of any governmental activity, we felt to be rather unsatisfactory. An independent and comprehensive Library service was, therefore, envisioned not merely as a repository of books, legislative debates and parliamentary papers, but to provide an up-to-date and objective storehouse of knowledge on which Members could draw with freedom and confidence.

The year 1950, when India became a sovereign, democratic Republic, saw the beginning of a systematic expansion of the Parliament Library's collections, both quantitatively and qualitatively. A humble start was also made in setting up a Members' Reference Service within the newly established Research and Reference Branch which initially functioned independent of the Parliament Library.

During the past four decades, the Library as well as research and reference services for Members have gradually developed into what is now familiarly known as LARRDIS (the Library and Reference, Research, Documentation and Information Service). The present organisational structure and nomenclature of the integrated service is the result of a major functional re-organisation of the Secretariats of the Lok Sabha and the Rajya Sabha brought about during 1974-1975.

The Setting: It may be mentioned here that the Constitution of India has adopted the parliamentary form of government and a federal set up. The national Legislature consists of two Houses-the House of the People (or Lok Sabha), and the Council of States (or Rajya Sabha). Direct elections to the House of the People are held every five years and the first past the post system is prevalent. The country is at present divided into 543 single member territorial constituencies for the purpose of elections to the Lok Sabha.

The Members of the Council of States are elected by Members of the Legislative Assemblies of the States. Seats in both Lok Sabha and Rajya Sabha are apportioned to different states in proportion to their population. With its sub-continental size and the huge population, it is only natural that the country's Parliament reflects a wide spectrum of cultures, languages, political preference, and so on. Members of Parliament, belonging as they do to diverse backgrounds, are in need of information and assistance on a wide range of subjects. The task before LARRDIS is, therefore, enormous as well as unique.

This uniqueness of LARRDIS can be seen from another angle. The Constitution provides for a separate Secretariat for each House of Parliament under a Secretary-General who will report to the Speaker of the Lok Sabha or the Chairman of the Rajya Sabha as the case maybe. The LARRDIS is an integral part of the administrative structure of the Lok Sabha Secretariat and is under the Speaker of the Lok Sabha. Yet all the information needs of Members of both Houses are attended to by this Service very much like the Congressional Research Service in the United States.

Library Committee: There is a Library Committee of Parliament consisting of six Members of the Lok Sabha, including the Deputy Speaker, and three Members of the Rajya Sabha nominated annually by the Speaker.

The Deputy Speaker is the ex officio Chairman of the Committee. The role of the committee is to advise the Speaker on all matters concerning the development of the Parliament Library. The Committee keeps a watch on the quality and quantum of all acquisitions of the Library and gives comments and suggestions on the running and functioning of the Library and its ancillary services. The Parliament Library in India is distinct from the national library. Whereas access to the national library is open to all, the Parliament Library is intended mainly for Members of Parliament. Former Members of Parliament also are eligible to make use of the Library under certain conditions. Research scholars in specified cases and assistants of Members are also allowed access to consult the documents though not to borrow books and other publications.

Organisation of LARRDIS: LARRDIS is presently divided into the following functional Divisions:

1. The Research and Reference Divisions
2. The Media and Research Division
3. The Parliamentary Affairs Division

4. The Library and Parliamentary Museum and Archives Division
5. The Research and Documentation Division, and
6. The Library Computerisation and Parliament Library Building Division.

The team of officers in each Division is led by a Joint Director. All the six Divisions are under the overall charge of a Director. Nomenclatures and the work allocated to these Divisions are subject to change, from time to time, depending upon the exigencies. At present we have a total of 154 research and library personnel working in the LARRDIS.

Parliament Library: The Parliament Library has holdings of about 1 million volumes, including debates of the Indian Parliament, the State Legislatures of India, and Foreign Parliaments, reports of Central and State Governments and the United Nations and its allied agencies, Gazettes of Central and State Governments, and other documents. The Library receives nearly 200 newspapers from different parts of the country and from around the world. Of these, 181 newspapers are from India itself, which includes 35 newspapers in English, 40 in Hindi and 106 in the different regional languages of India. Foreign newspapers are received from United Kingdom, USA, Pakistan, China, Singapore, Sri Lanka, Indonesia. Afghanistan, Malaysia and Nepal. The Parliament Library also receives regularly 842 periodicals of which 263 are from abroad. Out of the 263 foreign periodicals, the UN and its agencies account for 59 titles and the rest are from different countries. Of the remaining 579 Indian periodicals, 403 are in English, 94 in Hindi, and the remaining 82 are in various regional languages.

Regional Languages Books: Ours is a nation which abounds in diversity based on several factors, the main constant among them being the languages. Each one of our major languages is rich in its own way and in its distinctive literature. Together, these languages add to the richness of our Library heritage. Keeping this in mind, our Library has also laid emphasis on developing its Regional Languages collections. A separate wing, containing over 54,000 holdings on various Indian languages, has been carved out in the Library for the benefit of readers in regional languages.

Gandhiana: As a mark of respect to the memory of Mahatma Gandhi, the father of the Nation, and to make available all the works by and on the Mahatma at one place, there is a separate section called Gandhiana. This section contains about 1,700 books in Hindi, English and several other Indian regional languages.

Similarly, the Parliament Library also has a separate section exclusively devoted to the first Prime Minister and architect of modern India, Pandit Jawaharlal Nehru.

Staff Library: In order to enhance the reading habits of the employees of both the Lok Sabha and the Rajya Sabha Secretariats and their wards, a separate "staff library" has been set up. Books are added to the Staff Library, from time to time, as per suggestions of the staff. The present holdings of the Staff Library are about 22,000 volumes. It also receives 22 newspapers and 59 periodicals regularly. Nearly 60 new books are added to the Staff Library per month.

Rare Books: The Parliament Library has a fairly good collection of rare books. Books in this category number about 800. The oldest book available in the Library dates back to the year 1671.

Acquisition and Processing: The function of informing the Members about the new additions to the Library is primarily performed through our monthly publication Parliament Library Bulletin. Latest arrivals are also put on display on the "New Arrivals Counters". From time to time, book exhibitions are organised on various themes to mark specific occasions.

Addition of books to the Library is made through purchase, gifts and exchange. We have exchange arrangements with 50 countries, 74 foreign institutions/Universities and 129 semi-government organisations in India.

As mentioned earlier the Parliament Library is a depository of all UN publications. In all, we add 7,000 to 7,500 books and other publications to the Library every year. Of these 4,000 to 4,500 publications are purchased directly from the local booksellers. The acquisition procedure for books, particularly the initiation of titles for ordering, approval process, placing orders, accessioning, payments, acquisition order follow-up and on-line queries by readers, is now carried out using the computer software package

LIBSYS. We classify books according to the 20th edition of the Dewey Decimal Classification Scheme and catalogue cards are prepared as per the 2nd edition of the Anglo-American Cataloguing Rules. We are using the Library of Congress List of Subject Headings (12th edition) for assigning search key words. Catalogue cards are arranged according to their class numbers as well as by author and subject in alphabetical dictionary order. For our card catalogue we follow word by word arrangement.

Press Clipping Service: We have a Press Clipping Service which helps in the speedy disposal of references from Members of Parliament on current subjects. All important news and views, which include editorial comments and articles from 19 English and 8 Hindi dailies published in different States of India, are clipped and maintained in separate subject folders. Moreover, Press Information Bureau (PIB) releases, features of the Indian News and Feature Alliance (INFA) and other daily digests of news and views on matters if topical interest are scrutinised for being retained in the folders. These folders have been found very useful by Members and other researchers who are always asking for them. We are also planning to have the clippings microfilmed.

Reprography Service: A reprography service has also been set up to meet the urgent requirements of Members for photocopying important press clippings, parliamentary questions and answers, articles from periodicals and newspapers and extracts from books and other documents. We have four photocopying machines at present and Members of Parliament and accredited press correspondents can get free copies of the documents connected with parliamentary proceedings.

Documentation Service: One of the principal adjuncts of our Library is its Documentation Service. It indexes and prepares annotations of important books, reports and articles appearing in newspapers and periodicals received in the Parliament Library.

These indexes and annotations are classified and published in a fortnightly indexing periodical titled Parliamentary Documentation. They are also fed into the computer to facilitate searches. The service has now become an invaluable aid for research and reference activity. It also enables preparation of select documentation lists for the use of Members.

Microfilming Service: The Parliament Library has a large collection of old debates, magazines, reports, rare files, and other documents which are required to be preserved for reference over long periods. We have started to preserve these documents in the form of microfilm rolls. Our microfilming service is now equipped with microfilmers, cameras, processors, duplicators, and microfilm reader-cum-printers. There is also a Computer Aided Retrieval (CAR) system for retrieving information form the microfilmed documents.

Research and Information Service: The Research and Information Division of the LARRDIS regularly brings out books, monographs, brochures, background notes, and so on, on matters of parliamentary interest, including prospective legislative measures which are likely to generate demands from Members. Research notes are also prepared by this Service for use at Seminars, Conferences and Symposia and for the parliamentary delegations going abroad.

Our Research and Information Service brings out the following periodicals:

1. Journal of Parliamentary Information (Quarterly)
2. Digest of Central Acts (Quarterly)
3. Digest of Legislative and Constitutional Cases (Quarterly)
4. Abstracts of Books, Reports and Articles (Quarterly)
5. Diary of Political Events (Monthly)
6. Science and Technology-News Digest (Monthly)
7. IPG News Letter (Quarterly)
8. Public Undertakings-Digest of News and Views (Monthly)

Members' Reference Service: The Members' Reference Service is perhaps the one service which is the most sought after by Members especially during Session days. It collects and supplies, on demand, factual and objective information to Members of Parliament, Presiding Officers and Committees.

The Members' Reference Service also undertakes anticipatory referencing work and brings out Information Bulletins, Background Notes, and prepares Study Boxes on topical issues. For this, the requirement of Members is assessed well in advance and relevant material is kept handy. During the course of an average year the number of references received, and disposed of would come to about 5,000.

Computerised Information Service: The Parliament Library has introduced a computer-aided information system know as PARLIS. The database is created by the Library's Computer Centre with the help of the National Informatics Centre (NIC). Our Computer Centre is presently linked with the NIC's satellite-based network known as NICNET. It is possible to interact with various Departments/Ministries of the Central and the State Governments and other organisations with the help of NICNET. In addition the network links State Capitals and all the District Headquarters in India. The Parliament Library is thus able to exchange messages and other information with the State Legislatures and District Headquarters.

NICNET is also connected with major international networks like INTERNET, SPRININET, UUNET, TELNET, TYMNET, DATAPAK, and so on. Therefore, it is possible to inter-change information with foreign Legislatures too.

Members of Parliament can have access to computerised information/database available in the Computer Centre from their residences through their personal computers. The ultimate aim is to offer speedy and efficient service to Legislators through a well-articulated and versatile Library and Information Network.

Parliament of India WEB Page: The Parliament of India home page at the world wide web site of the INTERNET became active on 15th March 1996. The information contained in the documents of the Indian Parliament, which until now existed in printed form, is now available to people across political and geographical barriers. The Constitution of India, Constituent Assembly Debates, Rules of Procedure and Conduct of Business in Lok Sabha, Directions by the Speaker of Lok Sabha, Decisions from the Chair, Parliamentary Debates from November, 1995, and bio-data of Members of the Lok Sabha and the Rajya Sabha constitute the major components of the parliamentary information now available on the Internet. A comprehensive introduction to the constitution and functioning of the Parliament and facilities provided to the Members and to visitors is also given.

Audio-Visual Service: Computers, films, videos, and so on, have become normal elements of our Library culture and their products contribute to our knowledge and general world-view. Video News Magazines covering important national and international events in different areas of politics, economics, law, and so on, brought out regularly by various organisations and agencies, are of immense use to Members. Moreover, the archival value of recordings of parliamentary proceedings is also increasingly felt. Our Audio-Visual Service is maintaining a complete set of all the recordings of parliamentary proceedings from the time video recordings started in 1989. There is a proposal to acquire video copies of the proceedings of the Parliaments of other countries, too, in exchange for the cassettes produced by us. Language learning courses also form an integral part of the Audio-Visual Service.

Media Relations: LARRDIS caters to the needs of the press and other mass media, including television, radio, films, and so on, and provides all possible facilities to enable them to cover the day-to-day proceedings of the Lok Sabha. For this a separate Wing has been set up within the LARRDIS. It establishes liaison with the parliamentary press correspondents and the various government publicity organisations and communications media for the proper publicity of parliamentary and other activities of the Lok Sabha.

It displays on the closed circuit Television (CCTV) system, information pertaining to on-going business in both the Houses of Parliament. Important news items received through teleprinter machines are regularly collected, edited and displayed on news display boards at regular intervals throughout the day during session period for the information of Members.

Parliamentary Museum and Archives (PMA): The Parliament Library has also set up the Parliamentary Museum and Archives (PMA) which has a section of photographs and films to preserve an authentic, comprehensive, complete and up-to-date pictorial record of the history of the institution of Parliament, its activities and personalities. The PMA has a Museum Section which contains models, charts, objects and photographs depicting the evolution and functioning of parliamentary institutions in India and elsewhere. A separate Parliamentary Archives Section within PMA has also been carved out. It plans to acquire and preserve the records connected with the framing of India's Constitution, the growth and work of parliamentary institutions and private papers of parliamentarians, irrespective of their political affiliations. These records are open to scholars, private as well as official, for consultation to promote scholarly ventures in the sphere of parliamentary and democratic traditions, practices and institutions.

From time to time, the PMA organises exhibitions on varying themes mostly connected with the functioning and achievements of Parliament. All photographs, records, books, and archival documents are accessioned and their indexes are computerised for facilitating reference and research work.

Future Plans: As mentioned earlier, the Parliament Library which had a modest beginning in the early twenties, started growing with the attainment of independence by the country in 1947. Now it has a staff strength of more than 150. The collections of the Library, too, have grown rapidly and have crossed the million mark. For some time now, we have been finding it too difficult to manage with the existing accommodation both for books and for staff. At last a solution is in sight. A new Parliament Library Building is coming next to the Parliament House. The project is due for completion in 1998. The new Library Building with all modern facilities will help us in improving the information services, and creature comforts, too, available the users of the Library.

VOICE OF INDIA'S DISABLED: DEMANDING EQUALITY IN LIBRARY SERVICES

Introduction: A comprehensive country wide survey of persons with disability revealed that approx: over 90 million people are suffering from one or the other kind of disability. It can be said that about 12 million are blind, 28.5 are partially visually impaired, 12 million people have speech and hearing defects, 6 million are orthopaedically handicapped, 24 million are mentally retarded, 7.5 million are mentally ill and 1.1 are cured of leprosy.

The National Sample Survey Organisation (NSSO-1991) revealed that 1.9% of the country's total population were affected with physical and sensory disabilities and surveys conducted by various research organisations indicate that about 3% are mentally disabled. Disabled people in India have been subjected to direct and indirect discrimination for centuries. There is little doubt that it is discrimination not disability that disables people. They deserve a comprehensive legislation to tackle the blatant discrimination against them. With human rights organisations voicing their concerns both nationally and internationally and also acting as their guardians, an increasing number of countries are vigorously preparing policies favourable to the disabled people. In India the recent guidelines framed by the Govt. called "THE PERSONS WITH DISABILITIES (Equal opportunities, protection of rights and full participation) ACT 1995 ensures that the appropriate Govts and local authorities implement the Act which says:

a. ensure that every child with a disability has access to free education in an appropriate environment till he attains the age of 18 years

b. endeavour to promote the integration of students with disabilities in the normal school.

c. promote setting up special schools in Govt. and private sector for those in need of special education in such a manner that children with disabilities living in any part of the country have access to such schools.

d. endeavour to equip the special schools for children with disabilities with vocational training facilities.

Although the new Act will help to develop positive action programmes, there are no positive guidelines for helping the disabled to have access to an effective library and information services programme. If the law offers them equal opportunities they must also ensure that people serving the disabled such as the Govts and NGO's find new ways of serving the disabled. Here, Libraries and Library workers can play their part to hasten the process of full and total integration of the disabled in the society in a big way.

Needs of the Disabled: Individuals with disability have varying degrees of need. Their needs are just like those who are not disabled. They often strive hard for a high quality of life as other normal individuals. Unfortunately many a times people fail to understand that disability in simple terms is nothing but a natural part of the human experience. Often they are shrouded by misconceptions such as, that the disabled are forced to lead a poor quality of life.

But the fact is, a person with disability with all limitations can carry out normal activities of living if they have an easy access to community based long term services such as an attendant care, accesses to buildings, public transportation, side walks etc.. Even the severely disabled, when provided with quality health care services and the necessary equipments are able to carry on the tasks similar to those done by the non-disabled. One cannot but feel disturbed that inspite of having the capacity they have to continue remaining disabled because of lack of a strong community based support system. A disabled who is in the prime of his youth, demands an equal opportunity and must therefore be offered a range of assistance such as examination support, specialised equipments, library assistance, note taking in class, reader sign interpreters and parking provisions etc. Only when they have a strong support such as the ones mentioned above they can hope to lead normal lives.

Types of Disability: In order to assist the disabled in leading an active and purposeful life it is important to identify and understand the various types of disability that can be assisted. They can broadly be classified as Physical and Mental disabilities.

I. Physical Disabilities:

a. *Spinal Chord Injury:* It is an injury or damage done to the spinal chord due to an accident or a fall, which may result in partial or complete paralysis. The most common condition that occurs when there is a spinal injury is Paraplegia-meaning, paralysis of legs affecting both movements and sensation. Another condition is Quadriplegia-a degree of paralysis in all four extremities such as arms and legs.

b. *Amputated Limbs:* Amputees are those persons who have one or more missing limbs as a result of accidents caused by vehicles, machinery or any other deformity caused during birth. Such persons can be partially mobile.

c. *Blindness/ Visual Impairment:* There are several kinds of visual impairment of which the most common is blindness which can occur due to various reasons such as malnutrition during childhood, illness, or due to accidents. Persons with visual disability have a wide range of abilities as well as limitations. They may be able to read large print and may even move about without any mobility equipments in most situations or sometimes they may be able to perceive light and darkness and perhaps even colours.

d. *Hearing Impairment:* Individuals may be affected by hearing impairment at all ages, the extent of which could be mild or severe depending upon the age of onset, degree and type of loss. Each individuals adjustment to hearing loss is different from one another. Some can be assisted with hearing aids while others are not affected by it. Persons who are deaf due to hearing loss after the age of 20, tend to have fair understanding, near normal speech and hearing but may require instructions to gain useful speech reading.

II. Mental Disabilities

a. *Learning Disability:* It is a disorder which affects the basic psychological processes of understanding or using written or spoken language. It can damage the ability to speak, read, write, listen, spell or do mathematical calculations. Conditions such as brain injury, minimal brain dysfunction, dyslexia, and developmental aphasia are examples of learning disabilities.

b. *Mental Illnesses:* These encompass Schizophrenia, anxiety disorders and depressive disorders. Schizophrenia is a highly complex disorder which is caused due to a series of chemical changes in the brain. It usually occurs between the age groups of 15-25 years and is characterised by fragmented thoughts followed by an inability to process information. The condition affects the individuals family, professional and social life making him incapable of functioning normally. Surprizingly their intelligence is not affected and many of them are capable of leading partially normal life if they follow their regular pattern of medication and rehabilitation programmes such as those offered by half-way-homes.

c. *Mental Retardation:* It is distinguished from mental illness because of its presence before birth. A condition characterised by abnormal brain development in the womb not corresponding with normal physical growth. Their learning ability, reasoning power and judgement all develop at a slower pace. Accidents, poisoning, or illness after birth can be a cause for mental retardation. Many of the mentally retarded people are able to participate in activities with non-disabled people given an appropriate adaptation and support. Others may require a long term structured programme. With adequate training and education such persons can be more self reliant citizens. They can be found holding non-skilled or semi skilled jobs and can be made to effectively integrated into the social structure.

Creating a Model Library Programme for the Disabled: In order to create an efficient and effective library programme for the disabled, Libraries need managers who are upto date and are aware of the latest developments that are likely to have profound effect on their services. It is their collective responsibility to promote quality services by gaining a good insight into the problems faced by the disabled.

Library staff must recognise that some disabled persons have no control over their behaviour and must be competent enough to handle difficult situations. They must be prepared to give individual attention so as to understand their strongest communication mode. Therefore the following aspects are essential to develop a model library programme for them by way of: a) Training of Library staff, b) Developing User Assistance Schemes, c) Offering Special Services

a) *Training of Library Staff*: Professional library services depends largely on the continuous upgrading of staff through training on a regular basis. It could be fortnightly, monthly, or seasonal training. Special training requirements can be determined by the Library management and training Officers depending upon the skills and training needs of the individuals. Once the need has been assessed the method of training can be planned. Here is a simple and effective training plan that can be implemented:

Objective: To familiarize with important aspects of disability and disabled users in a Library environment.

Structure: A five day course of lectures, discussions and practical work. The practical training could be held in small groups giving each group an opportunity to work with all types of disabled users.

Materials: A resource package of the training manual.

b) *Developing User Assistance Schemes:* Once the Library worker gets trained they will feel that it is their collective responsibility to provide quality services by gaining a good insight into the problems faced by the disabled and thereby he/ she will be competent enough to create an efficient and effective user assistance scheme such as:

- Providing the disabled users with reading lists and catalogues, high demand materials and lecture tapes that are held in open reservation.
- Delivering books and other documents from library shelves.
- Photocopying of Library materials and enlarging for the partially blind.
- Paging books from stacks and shelf areas.
- Extending loan periods or modifying other lending rules on an individual basis.
- Accepting telephone requests and providing reference service.
- Conducting orientation tours and information skills sessions.

- Providing audio visual equipments.
- Assisting in the use of computer aided learning equipments such as CD-ROMS, optical disks etc.
- Providing access to library facilities by ensuring that directional signs are labelled at appropriate locations with large, simple and clear messages.

c) *Offering Special Services:* Different types of disability require different types of specialised services. The same theory holds good for library services as well and hence the trained library personnel will taken into consideration the following points to fulfil the objectives:

- Library services for the blind must enable them to have access to equipments such as Braille printers, Braille embosser and tape duplicators, Kuzweil reader (a text-to speed reading machine with synthesised speech output), closed circuit TVs for magnifying regular text, PCs with CD ROMS, Powermacs with CD ROMS, large print tape writers, special track tape recorders, computers that are having adaptable equipments such as voice eyes.
- Services for speech and hearing impaired users must include TTD communications (a device also known as TTY-text type telephone) for those needing library questions answered on telephone. Librarians must also familiarise themselves with American sign language which is considered to be the common language of the deaf community.
- For the physically challenged the library facility must be barrier free to wheel chairs and other mobility devices and ensuring that all devices including door handles are designed for easy manipulation.
- Persons suffering from mental disability can be served by helping to locate and retrieve materials. Because reading is a common problem for them they may enquire assistance in identifying the materials of both the print and electronic types of documents. They also need help in turning catalogue cards or keying in

commands on the computer. People suffering from mental illness such as Schizophrenia must be motivated to read books on self-development and may require constant counselling to use library resources to overcome their depression. This is termed as "Bibliotherapy", that is, using books for therapy.

Libraries and library workers can thus play a significant role in bringing hope to the otherwise dismal situations of the disabled world. Adequate financial support must be sought in order to sustain the above mentioned library services and subsequently expand them to suit the latest technological advancement. There is a need to explore new sources of funding taking into account the reality that the networking of Indian libraries can save funds by way of avoiding duplication of services and also help in providing an efficient inter library loan service thus reducing the time of the reader considerably. Like normal individuals the disabled also need information and other services to help them raise above their disability.

Library and information centres in educational institutions must be encouraged to raise funds by marketing their services. In the process they must try to integrate not only persons with disability with normal users but also integrate the electronic media with their collections and develop local networks. Libraries and Librarians will thus bridge the gap of disability and ability by creating a networked society where in, every individual will be connected without any discrimination.

DIGITAL LIBRARY OF INDIA

The recent advances in computer, storage and communication technologies are so stunning that it prompted one of the authors of this paper (Raj Reddy) to envision and explore the possibility of storing in digital form all the knowledge ever produced by the human race and making this content available free of charge to be browsed and searched by anyone, anywhere and at anytime. This vision is the goal of the Universal Digital Library Project (UDL). In addition, Reddy envisions a future where almost all information will be readable by humans as well as machines. The trend would be such that any information that is not on-line and accessible to search engines may become unusable. In a thousand years, only a few of the paper documents we have today will survive the ravages of deterioration, loss, and outright destruction. Hence there is an urgent need to preserve our knowledge and heritage in digital form. As a part of Raj Reddy's grand vision, a mission to digitize one million books was embarked upon as a collaborative project involving many countries, especially India, the United States and China. In this paper, we present the approach taken and the technological challenges associated with the Million Books to the Web Project (MBP).

An interesting offshoot of our efforts to make knowledge available free of charge to everyone is the opening up of research opportunities in language technologies – particularly for Indian languages – in order to make sure the language in which that knowledge exists does not become a barrier to information access.

Language technologies research in Indian languages has so far been impaired by the lack of resources pertaining to Indian languages in the form of text or speech corpora. Compared to the corpora in English and other European languages, or in Chinese, the resources available for Indian languages have been very limited. This situation is being changed by the Digital Library of India (DLI) initiative, which is the Indian part of the UDL and MBP.

DLI today has scanned over 289,000 books composed of approximately 105 million pages in the Indian and English languages. Digital representation and storage mechanisms have been developed for Indian languages, and a large number of applications are being built to store, process, retrieve and present the Indian language content.

The Digital Library of India fosters a large number of research activities pertaining to language technologies for Indian languages, and acts as a testbed for developments made in areas such as text summarization, information retrieval, machine translation and transliteration, optical character recognition, handwriting recognition, and natural language parsing and morphological analyses. We present here an overview of the activities of DLI in these areas and show how DLI is acting as a multilingual resource, even without the availability of manually curated data.

Vision: Digitization of All Human Knowledge

For the first time in history, technology seems to favour the possibility of digital preservation of all the significant literary, artistic, and scientific works of mankind, as well as the potential of free access to them from every corner of the world. A Universal Digital Library (UDL) has the potential of improving the global society in ways beyond measurement. The Internet can house a Universal Library that is freely accessible to everyone. This would revolutionize education for all our future generations. There were about 10 million unique book and document editions before the year 1900, and about 100 million since the beginning of recorded history. An average-sized book is around 250 pages and would require about 50 MB of disk storage if the book were stored as compressed images.

Thus, all the books and documents ever produced by the human race would require 5 peta byte of storage. Even if we multiply this by a factor of 200 for all other forms of knowledge, such as music, images, audio and video, the total of that information could be stored in a zeta byte server.

With the storage capacity of digital disks increasing by a factor of 1,000 in ten years, it looks technically feasible and financially affordable to articulate the vision to store on the computer all forms of knowledge ever produced by the human race. With new digital technology, though, this task is within the reach of a single concerted effort for the public good, and this effort can be distributed to libraries, museums, and other groups in every country. This formed the motivation for the grand vision of the Universal Digital Library Project.

Mission: A Million Books on the Web by 2008

It is believed by some that the goal of creating a Universal Digital Library is impossible and that attempting to create it could take hundreds of years and still may never be completed. Nevertheless, as a first step toward realizing this grand vision, a project was proposed that would create a Universal Library starting with a free-to-read, searchable collection of one million books available to everyone over the Internet by the year 2008. This first major project toward building a Universal Library is named the

Million Book Digital Library Project (MBP). Within 10 years, it is expected that the collection will grow to 10 Million books.

The result will be a unique resource accessible to anyone in the world, 24x7, without regard to nationality or socioeconomic background. Typical large high school libraries house fewer than 30,000 volumes. Most libraries in the world have fewer than a million volumes. The total number of different titles indexed in OCLC's WorldCat is about 55 million. A library of one million books, therefore, contains more than the holdings of most high schools, and is equivalent in number of volumes to the number in the libraries at many universities, representing a useful fraction of all available books.

The MBP is a multi-country project, and that part of the project being performed in India is called the Digital Library of India (DLI). A secondary objective of the MBP project, and hence the country-specific Digital Library exercises such as DLI, will be to provide a testbed that will support other researchers who are working on improving the technology required for scanning and indexing. The corpus the MBP project creates will be one to three orders of magnitude larger than any existing free resource. It is expected that when DLI matures, more than 10,000 books will be available in each of the major Indian Languages. Such an immense and unequaled volume of data would also provide an excellent resource for language processing research in areas such as optical character recognition (OCR), machine translation, summarization, intelligent indexing, and information retrieval.

Collaboration

Accessing, scanning and web hosting a million books is a major logistical challenge. A million books can be expected to comprise approximately 250 million pages. Throughput of a scanner used in the project (Minolta PS7000) is about 6,000 to 10,000 pages per day, based on three-shift operation. Using 100 scanners, and a 1,000-person workforce, the task of scanning a million books could be completed in about 420 working days, or roughly two years of three-shift operation. However, given the uncertainties of having skilled workers available at all times, and experiencing equipment failures and power outages, it is more reasonable to expect that the scanning would be completed in around four years.

The United States of America, India and China are currently the major collaborators in the MBP project with regard to technology development, operations and management. Collaboration among these countries has been a factor in the project's success so far, because these three countries provide access to a number of large academic libraries, of which over 700 are OCLC members from India and China. Also, since the scanning centres can be established anywhere, UDL has chosen to establish them close to these academic libraries so that the books can be transported with minimal difficulty.

The collaboration so far has been quite fruitful, with more than a few thousand books digitized every month from all the various scanning points established in India and the US. Currently, we have more than 200,000 books scanned and ready for use.

The Approach: Digital Library of India

The Indian Institute of Science (IISc), Carnegie Mellon University (CMU), the International Institute of Information Technology, Hyderabad (IIITH) and many other academic, religious and government organizations, totaling about 21 "Content Creation Centres", have become partners in the Digital Library of India (DLI) initiative for the digitization and preservation of Indian heritage present in the form of books, manuscripts, art and music. Each centre brings its own unique collection of literature into the digital library.

Many authors have cooperated by contributing their books to the digital library and making them available free of charge to anyone. This digital library is also intended to be a testbed for Indian language Research. DLI is intended to be a leading and contributing partner in worldwide efforts toward making knowledge free. A pilot project to scan around 10,000 books was initiated at CMU and then followed up at IISc, IIITH and other organizations; all the processes involved have been perfected.

The vision is to use the disruptive technologies like the ICT to preserve all the knowledge of the human race in digital form and make that content searchable, independent of language and location, and to ensure that the rich cultural heritage of countries like India is not lost during the transition from paper to bits and bytes, as they were lost during a former transition of cultural content from palm leaves to paper.

Today, most of the works created by humans – be they in the form of books or music or movies – are "born digital". Hence if we build the digital library with a proper framework and architecture, it should become possible to bring the library up to date more easily than it was to digitize physical formats. But for the time being, we are concentrating on digitizing these physical formats.

For the digitization of old books, we have developed a complete process comprised of scanning with planetary scanners, cropping, cleaning up images, and using software for the OCR conversion of English documents, format conversions and search engines. With available technologies, a 500-page paper book can be digitized and made available on the web in about two hours without having to unbind the book. So far, more than 289,000 books have been scanned, of which nearly 170,000 are in Indian languages. More than 84,000 books (25 million pages) are available on the DLI web site hosted by the Indian Institute of Science (http://www.new.dli.ernet.in), and more than 149,000 books (43 million pages) are available on the DLI web site, which is hosted by the International Institute of Information Technology (http://dli.iiit.ac.in).

The link to other partner sites are also provided through (http://www.new.dli.ernet.in). Contents between the two sites overlap in order to ensure fail safe availability. The books can be accessed from either of these web sites. With the success of the joint efforts of the IISc-CMU collaboration, many other nations including China and Egypt have shown interest in participating in this effort, making it truly a global effort in knowledge sharing. China has made significant progress and has taken the Million Books to the Web as a National Initiative.

4

TECHNOLOGICAL CHALLENGES

QUALITY MANAGEMENT

In large digitization projects, with collaborative work and distributed efforts of various parties involved in the process, we often find a compromise of quality, which enables errors to creep in. As is the case in any digitization project, DLI work is performed by humans and machines. Hence, occasional errors are possible, and these can be broadly classified into two categories: human errors and machine errors, both of which we discuss below.

Human Errors

Human errors in the digitization process are perhaps the costliest of the errors; they arise most often due to miscommunication between project staff or to staff incompetence or non-adherence to process and standards. Most of the books scanned in the DLI project are procured from sources like libraries and government archives, and the records for these books contain metadata entered by knowledgeable personnel. Although, in general, the metadata can be relied upon, the quality of that metadata are nevertheless subject to individual biases. In addition, for a major portion of books scanned in the project, accompanying metadata exists only in non-digital formats, and these have to be entered manually. For the data entry of metadata of a book, we largely rely on librarians to assess the accuracy and credibility of that metadata. But this situation is not foolproof.

For example, a particular librarian might not be well advised about the hierarchy and ontology of book classification and so might classify a book as belonging to the category of "Art" when it really belongs in the category of "Music". Consequently, the misclassified book would not show up in search results for the end-user who is interested in and searches for "Art", and instead would show up as a non-relevant search result to a person searching for "Music". An even worse situation would arise if the book were erroneously assigned to a completely irrelevant category.

Similarly, the process of scanning and producing digital content needs to be done with proficiency and care. An improper scanning operation made without following the standards set, may result in a digital collection that is not useful or suitable for an end-user. These problems manifest in various forms like the page of a book slipping while being scanning or an incomplete scan of a page. Manual errors can incur significant costs to the project, as firstly there is no satisfactory way to identify such errors when data is generated on such a massive scale, and secondly the erroneous data generated will not useful to end-users, thus undermining the purpose of the project.

Machine Errors

Machine errors are those that creep in due to the understandable limitations of the software being used or from improper configurations of the machines and software. Data generation, which is done during the scanning, is the key phase of the digitization process as we obtain digital images from the books. Once these images are obtained, the book is sent back to the library from which it came, and hence any problems that require that the book be re-obtained and re-scanned are costly errors.

For example, an image processing algorithm that checks for the 'skew' aspect in images may have certain limitations for detecting it and could classify a non-skew image as containing skew. Such a problem could be rectified with better versions of the software whenever available. However errors due to improper configuration of scanners during the scanning phase are more serious, as the data generated becomes less useful due to its low quality. Optical Character Recognition (OCR) is the software module that reads an image, understands the textual content in the image and outputs the text. Text is the appropriate means for storing data, occupies less storage than images do, is easily editable, and helps in the indexing and searching of the documents. OCR is thus a very crucial aspect of the process of creating the digital library. However, OCR is not 100% accurate, due to the various limitations imposed by the underlying recognizer. Let us assume that a page consists of 30 lines, each containing 30 words, and that a word, on average, consists of 4 characters. In that case, a typical page would contain 3600 characters. If the OCR is 99% accurate, it will result in 36 erroneously converted characters per page, and therefore for a book of 500 pages there would be about 18,000 errors in the OCR-generated text. Such a situation would be unacceptable to the end-user, although it does not significantly affect the efficiency of indexing and searching.

DATA MANAGEMENT

Scalable and Sustainable Architecture: Assembling the data and making it available for easy access is one of the most important phases of any digitization project. Each Mega scanning centre is responsible for gathering the metadata and the scanned content from the contractors operating at the scanning locations. This data is to be enabled on the web and also preserved for future. Enabling many tera-bytes of data for access to everyone in a highly reliable manner is needed for the success of the efforts put into the digitization process. Also data synchronization and management across centres needs to be done to reduce duplication and ensure reliable high availability and immediate recovery in the event of storage media failures and server failures. Finally, digital preservation of the collections for a long time into the future remains a very significant problem faced by any digital library.

Preservation Management: The books scanned for the DLI project are for the use of everyone for the foreseeable future. Hence preserving the content of these books is important. In addition, the data being frequently commuted between the centres needs to be preserved uniquely to ensure easy workflow management. Every book that is scanned and stored is associated with a unique barcode and descriptive metadata for identification, search and retrieval.

Synchronization Across Different Centres: Because the books to be scanned come from various sources – like libraries, government organizations, institutions and personal collections – that are distributed across the country, there could be duplicates among scanning locations maintained by a Regional Mega Scanning Centre (RMSC) and also across different RMSCs. However, the project cannot afford the extra cost of scanning these duplicate books, processing their images, and performing quality assurance on them.

Thus, communicating metadata across centres and within scanning locations is important. The duplicate books can be identified only by using metadata of a book like the title, author, publishing year, edition, etc. However, if the metadata is incorrect, missing or incomplete, as discussed in the previous section, it makes the duplicate detection all the more difficult.

Rich Metadata The metadata formats that are traditionally used for physical books, though comprehensive, are not sufficient for handling digital objects. Hence, we have had several discussions in the UDL and DLI projects about identifying the metadata that should be preserved along with the digital objects, and we finally narrowed down our requirements to the following three sub-categories of metadata:

1. *Regular Metadata:* Regular metadata contains information about the book like title, author, date of publication, publisher, ISBN, keywords, subject, language etc. We follow the widely understood and accepted Dublin Core metadata format, extended with a few fields like edition information of the book, and the use of OM transliteration for Indian language texts developed for DLI.
2. *Administrative Metadata*: Administrative details of the book, like the location where the book was scanned, the original source of the book, scanning details, etc., may not be of interest to the end-users of the book but are useful to the operational organization. These details can be used to trace the progress of the project, generate reports and identify bottlenecks in the scanning process. For example, it would enable us to trace the scanner that was producing low quality scans.

3. *Structural Metadata*: We have adapted the structural metadata concept for a book object in our digital library. This metadata contains information pertaining to each page, like the size of each page and whether that page is blank or has an important context attached to it – such as the beginning of chapter, end of chapter, index, preface, table of contents, etc. Such information enables us to improve the navigation of the book for the end-user and also improves search and retrieval systems.

Replication of Storage: Preservation of digital data is known to be a hard problem to solve. We attempt to address this problem to the best possible extent, by replicating the resources. All the books scanned in the DLI project are replicated and preserved in different locations across the world. We also store the content in two different formats – DVD/CD and hard drives. We are currently planning to have 3 different locations across the world (CMU, IISc and IIITH) where all the data scanned and digitized has been preserved in the above mentioned formats.

Architecture of the ULIB Project

In this section we describe the architecture that supports the process and workflow discussed in earlier sections of this paper. The architecture of the DLI project is motivated by factors like scalability, ease of maintenance, dependability and economy. All the tools and technologies used by DLI are open source.

Many issues related to interoperability and collaboration arise due to the huge number of books in different languages that are scanned at the various scanning locations, and the differences in the infrastructures used to preserve these digital objects. We solve these issues by deploying a distributed, decentralized architecture for the DLI project and by modularizing the tasks, using technologies like XML, databases, web services, etc.

Below, we first describe the architecture of the DLI portal hosted at each Mega centre (DLI-RMSC), and we then propose an architecture for organizing these individual portals in a decentralized and service-oriented manner to ensure a highly available and dependable DLI system.

Architecture of Mini UDL and DLI Hosted at Mega Centre

Each centre hosts the books that are scanned in the locations maintained by it. Currently there are three operational mega centres.

The digital objects produced by the scanning are preserved on Terabyte servers, which are clustered as a data farm. Each server in the data cluster hosts all the digital objects preserved on it, through an Apache web server. The cluster is powered by Linux and enhanced by LTSP3, an add-on package for Linux that supports diskless network booting.

This option of diskless network booting helps us boot a server without having to devote any space for storing the system-specific and operating system files. This set up is economical and also easy to manage, such that we can add or replace data nodes in the cluster instantaneously without the need for operating system installations and configurations. We have customized the kernel in LTSP to support hard disk recognition and usb hotplug, and to run a lightweight Apache web server.

As shown in Figure, the 'Linux Loader' machine runs a copy of this distribution of the Linux with LTSP. Each data server in the data cluster downloads the kernel over the private intranet and boots from it. The servers implement a hardware based RAID to contain disk failures, which adds to the reliability of the system. In addition, for data restoration in the event of an irrecoverable crash, a redundant copy of the complete data is present on external storage media.

The 'metadata server' is a repository of the complete metadata, which is in XML. XML has been chosen for its important role in enabling interoperability. Metadata is passed on constantly between contractors and the RMSC, and it also acts as an identifier of the book that is to be scanned. Using XML as the format modularizes the work by decoupling the RMSC and contractors, and it also ensures smooth interoperability. Wrappers present on the metadata server automatically populate the database from the XML metadata. Along with the metadata of the book, the database also contains pointers to the location of the book in the data cluster. The portal has a front end user interface that a user can log onto, use to query the metadata, and retrieve books he or she wishes to read online.

A caching mechanism deployed on the metadata server helps us cache similar queries posed to the database and return the results promptly. When a user requests viewing the complete content of the book, the location of the book in the data cluster is gathered from the database, the content is retrieved over http requests from the particular server in the cluster and then is broadcast to the user.

The 'proxy servers layer' between the Data cluster and the DLI-RMSC portal also has a caching mechanism enabled that handles repeated requests for the book pages, and this ensures quick response times. Like the metadata, books are preserved in text format, which makes them searchable. Currently, search is limited only to books in the English language, due to unavailability of optical character recognizers for other languages. The search is supported by Lucene.

Distributed Architecture of DLI Projects

The scanning operations of DLI take place at different locations in which the RMSC operates, and the digital data from its region is accumulated to be hosted online. DLI as such follows a distributed and decentralized architecture with each RMSC as an independently operating node. Decentralized architectures by definition avoid having central points, as they are candidate single points of failure and a performance bottleneck.

However, since the digitization process is a cumbersome process, the data that is the end product of the process is very sacred. Hence, redundancy is always advised in a data centric project like DLI, and therefore every RMSC that is a node in the decentralized architecture of DLI hosts the complete data from the other nodes. Currently, synchronization of complete book content between nodes is by physical transfer of Terabytes of information between the RMSCs.

We propose a Service-Orientated Architecture (SOA) for smooth interaction between the nodes in the DLA decentralized architecture. The following are a few advantages of the decentralized SOA of the DLI:

- Web services address issues of interoperability that arise due to varying media, databases, languages, middleware and operating systems across RMSCs.
- Metadata of books is synchronized between RMSCs via web services, on a periodic basis. This also helps in duplication verification across RMSCs.
- Other specific features like copyright information verification, statistical reports, etc. can also be exposed by the RMSCs and can be utilized across the DLI project via web services.

A user can log into the central site and request reading books online, at which point he is redirected to one of the closest RMSCs and will be served from that RMSC. This ensures quick response times for the user and also reduces, to some extent, load on any one set of servers.

Indian Language Technology Research under the DLI Project

The Million Books to the Web Project (MBP) acts today as a catalyst for research in Indian Languages. The primary goal of the DLI project, apart from making the books available online, is to make them available in fully functional form. The DLI is not simply a static repository of books – it has made possible bringing home the language and information processing technologies for Indian languages. In the rest of the paper we describe some of the research activities progressing under the umbrella of DLI.

Major impact contributions of the Digital Library of India to the Indian language information technologies are:

1. OCR for Indian languages
2. OM transliteration, which is an integral component of all the other systems
3. a text editor for Indian languages that is available for everyone
4. the book-reader interface that extends the availability of the Digital Library from *"anyone, anytime, anywhere"* to *"any-language"*
5. a machine translation system that we call *Good-Enough Translation* (GET-across) and

6. a search engine for Indian language texts.

These are briefly described below.

OCR in Indian Languages-Kannada

Designing an accurate OCR in the Indian languages is one of the greatest challenges in computer science. Unlike European languages, Indian languages have more than 300 characters to distinguish, a task that is an order of magnitude greater than distinguishing 26 characters. This also means that the training set needed is significantly larger for Indian languages.

It is estimated that at least a ten million-word corpus would be needed in any font to recognize Indian languages with an acceptable level of accuracy. DLI is expected to provide such a phenomenally large amount of data for training and testing of OCRs in Indian Languages. Many of the contents, besides scanned images, have been manually entered for this purpose. Using this extremely large repertoire of data, a Kannada OCR has been developed.

Optical character recognition for Kannada: The first block of the OCR is the segmentation algorithm that segments lines, characters and, within the character, a 32 X 32 block to identify the different key strokes that make up the character. These take care of the morphological dilation, base character, vowel modifiers and consonant conjuncts. Base characters are then normalized to 32 X 32, and the consonant conjuncts, as well as modifiers, are resized to a 16 X 16 matrix. Through a series of signal processing algorithms using DCT and KLT, the features are extracted. Structural features include aspect ratio, stroke at different orientations, and height of the segment, in the top zone, and the width of the character in the middle zone.

A neural network-based classifier is then used for training with the extracted feature vectors and testing. The current level of accuracy that we get is around 96-97% on clean documents scanned at 400 dots per inch. This accuracy falls to 40-50% if the image is of bad quality. Efforts are underway to further improve the accuracy of the OCR with better segmentation, faster speeds and enhanced training. This OCR is currently being extended to other Indian languages, including Tamil.

Om Transliteration: Unified Representation for Indian Languages

India is fast becoming a software superpower; however, PC penetration is merely 1.4% compared to that of television (17%) and telephone (5%). One of the limiting factors in low PC usage is the unavailability of the operational software in native Indian languages, and the language barriers between people. While the development of an operating system in a native language is one solution, this solution is likely to be limited to only a few languages. If the Indian language texts were instead available in parsable English-like texts, they would benefit from the advances in the language processing of other international languages. Isolated development of digital representations for the different Indian languages may further widen the language barrier in the country.

Thus there is a need for the development of a digital representation that lays a common foundation for all Indian languages. For seamless adaptation of algorithms in language technologies, this representation must also be parsable by universal language processing tools and algorithms, such as for machine translation, information retrieval, text summarization and statistical language modelling.

The representation must exploit the similarity in the alphabet of the various Indian languages. A large number of Indians are bilingual; while they can freely read and write in the one Indian language that is their mother tongue, and English, they can understand many other Indian languages as well due to the similarity of the origin of words. With this backdrop, we first developed a representation scheme for Indian languages, called *OM transliteration,* which formed the basis for all other work in Indian language research at DLI.

Om uses the same representation for keyboard input and formation, and digital storage. It is similar to ITRANS in that it uses combinations of letters from the English alphabet to represent Indian language syllables. ITRANS is a representation of the Indian language alphabet in terms of ASCII.

However, the OM transliteration developed under DLI is case-independent, and avoids excessive use of non-alphabetic characters; where used, they are consistent. Furthermore, the English alphabet combinations are designed such that they are easy to remember at the time of input using a standard keyboard, and they are also natural to read like English.

The case-independent representation allows the use of sentence and title case writing in a natural fashion; in addition, the texts are more highly readable than their ITRANS counterparts. It may be noted from any ITRANS text that the large mixture of capital and small letters and non-alphabetic characters leaves ITRANS text highly difficult to read. Om's features enhance usability and readability; it has been designed on the following principles:

1. easy readability,
2. case-insensitive mapping (While preserving readability, this feature allows the use of standard natural language processing tools for parsing and information retrieval to be directly applied to the Indian language text.), and
3. phonetic mapping, as much as is possible. (This makes it easier for the user to remember the key combinations for different Indian characters.)

ASCII representation may be used simply as a means of typing the text with a standard keyboard. For transliteration to Indian languages, OM representation is mapped to the Indian language fonts for display or converted to any other format, such as Unicode, where required. When a user is not interested in installing language components, or when the user cannot read native language script, the text may be read in the English transliteration itself.

Because India is a multi-lingual country with an inter-mixed population, often the Indian people can speak and understand more than one Indian language as well as English. Hence even in the absence of OM to native font converters, people around the globe can type and publish texts in the OM scheme that can be read and understood by many others, even when they cannot read native script. The readability criterion that has benefited from the case-insensitive phonetic mapping thus proves very useful. The major contribution of OM is to separate storing and rendering, which makes it language-independent across the Indian languages.

Om Text-editor

An integrated transliteration package that accepts OM ASCII keystrokes as input and maps them to native fonts has been developed. The script in any one of the supported true type fonts is sent to MS Winword for further formatting and layout options. Since the OM scheme is common to all the Indian languages, the display of the text can be converted between the supported languages by choosing it on the menu. The tool also integrates with email clients on the windows platform. A web interface with similar functionality has also been developed. The text may be saved as OM (ASCII) text, native-font text or in Unicode.

The tools have been used extensively for data entry for texts that feed into applications such as machine translation and optical character recognition. It has also been used purely for content creation by the outside community. An example may be seen at the magazine section of <http://www.telugumn.org>, where the story of Ramayanam has been created using this software. The integrated editor will be provided for hosting or use at any website free of cost, such as has been done at <http://www.telugumn.org>. The integrated editor is available for both windows and linux platforms. For those who wish to create content using a web interface, without the need to install the package, a java-based web interface is also available.

Multilingual Book Reader Interface

The books on the digital library of India are available to *anyone, anytime, anywhere*. The goal of this work was to add to it the dimension *of any-language*. While OM transliteration helps one to read the text of one language with a script of another, it does not provide any translation. Due to the grammatical and etymological similarity amongst Indian languages, and their phonetic similarity, OM takes things a step beyond mere transliteration. The understanding can be improved further by the simple technique of merely translating some of the frequently occurring words from corpora.

This has been the motivation behind the development of a multilingual book reader that supports *automatic transliteration* and *word to word translation* between Indian languages and between an Indian language and English. A Universal Dictionary, with the objective of providing the digital library user with a "good enough comprehension" of the content of the books in languages other than his own, has been built. This Universal Dictionary, which is cross-lingual, currently contains six Indian Languages (Hindi, Telugu, Assamese, Tamil, Kannada and Malyalam) besides most of the European languages.

When presented with electronic text in any Indian language, the book-reader allows text to be transliterated into any one of the many Indian languages. This is made possible with the OM transliteration scheme discussed above. This allows the user to read, for example, Hindi text in Telugu font. With the help of the Universal Dictionary, a word-to-word look-up table translation is made on the Indian language text between any pair of the many Indian languages supported by the Universal Dictionary. When a word is not found in look-up table, only its transliteration is displayed. While Indian languages are phonetic languages, English is not phonetic. In order to display English words in native language where required, a pronunciation dictionary is used.

These features provided by the interface are desirable not only to the readers who can understand but not read their own language, but also to those who desire to obtain at least a crude translation of a book to their desired language. The book-reader performs the functions involving transliteration independently while also connecting to the example-based machine translation system on the backend for full-text translation. The reader, while especially suited to a multilingual country like India, is also extendable to any other digital library where the resources of translation and transliteration are available at large. The multilingual book-reader presents novel features that improve the usability and reach of any digital library.

Indian language Search Engine-Tamil Search Engine (OmSe)

Technology for the deployment of information retrieval in Indian languages has been demonstrated by the development of the OmSe search engine using the off-the-shelf open source software *Greenstone search engine.* The use of the OM transliteration scheme makes it possible to store any Indian language information in ASCII and to use any conventional search engine for information search and retrieval. For illustrative purposes, Tamil documents stored in the ASCII representation of OM have been built and are directly available for indexing, search and retrieval without any modifications to the text-handling modules of the search engine.

At the time of display, the retrieved text, in addition being made available in readable English transliteration, is also converted to native Tamil script and displayed. Currently commercial quality optical recognition software is not yet available for Tamil and other Indian languages. Hence to demonstrate the technology, the Tamil search engine is built over a collection of born-digital newspaper articles, crawled from the web. The basic architecture of the Tamil search engine includes a server that contains a database, a web crawler that crawls and downloads Tamil-language web content from various Tamil web portals, and the OM transliterator that converts true type font ASCII to OM transliteration format.

The front-end of the search engine is the client side, which has a graphical user interface that prompts the user to type in the search query in OM transliteration format. The query typed by the user is also displayed in Tamil font to enable the user to make corrections, if required, while entering the keyword in OM Transliteration format.

The interface between the client side and the server side consists of matching the user query with the entries in the database and retrieving the matched web pages to the user's machine. The search engine takes the query to the database and looks at the matches as per ranking. The search engine then sorts these database entries using a ranking algorithm. Greenstone's ranking algorithm determines the relevancy of a retrieved webpage to the user query. The retrieved sites are then displayed along with links to these sites in text format.

Machine Translation

Example Based Machine Translation (EBMT) is basically translation by analogy. An EBMT system requires a set of sentences in the source language and their corresponding translation in the target language. A bilingual dictionary comprising of Sentence

Dictionary, Phrases Dictionary, Words Dictionary and Phonetic Dictionary is used for the machine translation. Each of the above dictionaries contains parallel corpora of sentence, phrases and words, and phonetic mappings of words in their respective files. These files would form the database for the machine translation system.

The basic premise is that, if a previously translated sentence occurs again, the same translation is likely to occur again. However, it is not possible to store the database as a set of a huge number of sentences. Instead, we can store the frequently occurring phrases and their translations and use these translations to translate a complex sentence. A sentence can be seen as a combination of phrases. Each sentence can be divided into a set of phrases and words. Instead of storing the entire sentence translation in the database, it would be more efficient to store phrase translations and word translations. This would optimize the database required for translations.

The EBMT has a set of 75,000 of the most commonly spoken sentences originally available in English. These sentences have been manually translated into three of the target Indian languages – namely Hindi, Kannada and Tamil. Bilingual word and phrase dictionaries between these target languages and English of over 25,000 and 18,000 entries respectively, were also created manually. The artificial intelligence engine learns from these examples and provides a good enough translation by looking for the longest match at the sentence, phrase and word levels. An even better way of storing the database is to store the rules related to the language pair. This has been observed to improve results dramatically. Though we would say that it is EBMT, the rules database would drastically reduce the size of the database and would actually improve the translation results.

Thus the phrase translations and phrase rules play a significant role in machine translation. The advantage of this simple "good enough translation" system is that its performance can be improved almost linearly with the increasing corpus and rule base.

The web enabled version of the EBMT is available at <http://bharani.dli.ernet.in/ebmt/>. The current machine translation system supports the following language-pair translations:

1. English to Hindi
2. English to Kannada
3. English to Tamil
4. Kannada to Tamil

The EBMT also has an interface for learning rules from its users. User feedback is stored separately and used later, after a verification for correctness, so that the main data base, and hence the performance of the machine translation system, is constantly improving.

Conclusion

The Digital Library of India (DLI), besides providing an opportunity to create freely browsable and searchable documents of value to all of humanity, has become a testbed for Indian language research. DLI has developed search engines, and clusters for data base storage and retrieval. It has also helped in creating a tight bond for those doing research in India within 21 centres, spanning academia, government and religious institutions. It is heartening to see that in India the Government, policy makers, religious institutions and scientists have put India on the world map of digital libraries and have done so as a national project under the Ministry of Information and Communication Technology. Many organizations are waiting to join this national initiative.

DIGITAL LIBRARY PROJECTS IN TAIWAN

The impact on human societies of various fast-developing information technologies has been studied for a long time. These studies have led us to believe that we are now facing critical cultural and social changes – changes in the means of communication from printed materials to electronic documents, and changes in lifestyle. Information technologies provide people many opportunities to digitize cultural treasures and heritage objects so they can be preserved and utilized in the digital era.

Otherwise, some of them might gradually become inaccessible or lost entirely. Realizing the importance of digital content, the National Science Council (NSC) of Taiwan initiated pilot projects in 1998 to study the feasibility of digitizing national cultural holdings. The first such project was the Digital Museum Project (DMP), which lasted from 1998 to 2001. Dozens of universities and research organizations participated in this project.

After three phases of the DMP had been completed, a five-year National Digital Archives Programme (NDAP) was launched in 2002 with support from the NSC. The main goal of this national programme has been to promote and coordinate content digitization and preservation in leading museums, archives, universities, research institutes, and other organizational content holders in Taiwan. The NDAP is also a sub-project of e-Taiwan, which belongs to a higher-level national development project in Taiwan.

In May 2002, the Executive Yuan of Taiwan proposed "Challenge 2008", a comprehensive six-year national development plan to foster creativity and cultivate talent among the Taiwanese people and to transform Taiwan into a "green silicon island". The plan is estimated to cost NT$2.6 trillion (approximately US$75 billion). Among the ten identified major development areas in the "Challenge 2008" plan is the construction of Digital Taiwan (e-Taiwan).

The e-Taiwan project aims to adopt information and communication technologies for developing Taiwan into a new knowledge-based economy, upgrading the competitive advantages of local industries, establishing a highly efficient government, and promoting a high-quality information society. This paper focuses on the development of the DMP and NDAP, including their background, organization and functions, and outcomes, and on the research-related issues of digital libraries. This paper also describes some of the obstacles and challenges facing digital libraries that may influence the effectiveness of digital library development in Taiwan.

Digital Museum Project

The main goal of the Digital Museum Project (DMP) was to integrate and establish several topical or technology-based digital museums that focus on disseminating cultural and educational content via information technology to the general public and students in Taiwan. The DMP project was divided into three phases.

- Phase 1 (September 1998-August 1999) was composed of two types of projects: topic-based projects (digital museums on various subject topics), and technology support projects. The topic-based projects in the first year included local or traditional culture and natural science subjects. For example, "The Discovery of Tamsui River" and "Indigenous Taiwanese – The Pingpu People" were about local Taiwanese culture; "Taiwanese Butterflies" and "Native Plants and Fishes of Taiwan" were about natural science and environmental ecology. There were also three digital museums projects on traditional Chinese culture: "Traditional Thoughts and Literature (The Four Books, Lou-Chuang, Poems of the Tang Dynasty)", "An Immortal Palace – Han Dynasty Culture and Burials", and "Firearms and Warfare of the Ming and Ching Dynasties". Technology support projects in Phase 1 consisted of five different areas: geographic information systems, word net, metadata, digital rights management, and digital library evaluation.
- Phase 2 (January 2000 to December 2000) was announced to the general research community through a request-for-proposal invitation. It drew wide attention and attracted nearly 90 topic-based proposals, from which 12 proposals were selected and funded, including four continuations from the first phase of the DMP. During the second phase, two technology support projects, one on a metadata management system and the other on digital watermarking, were carried out. Since the second phase projects were open to everyone in academia, the principal investigators of the proposals were affiliated with many universities and organizations with valuable cultural heritage.
- Phase 3 (2001) of the DMP included 15 topic-based projects of a wide variety, such as language literature, religious art, folk culture, historical relics, mathematics, biology, architecture, and geography. The outcome of the DMP

was quite fruitful; however, the scale and budget of the DMP (only $130 million NTD, ca. US$4 million) was not adequate to cover the entire cultural heritage of Taiwan. Hence, in 2002 the NSC proposed a new programme, the National Digital Archives Programme (NDAP), as a sub-project of e-Taiwan.

Background of the National Digital Archives Programme

The first phase of the NDAP (http://www.ndap.org.tw) was launched on January 1, 2002, and is a 5-year programme spanning the period from 2002 to 2006. This programme, also sponsored by the NSC, is to promote and coordinate content digitization and preservation at leading museums, archives, universities, research institutes, and other major organizational content holders in Taiwan. The goals of the NDAP are designed to accomplish the ultimate mission of preserving cultural heritage and sustaining cultural development activities in Taiwan. The goals are to:

1. Preserve national cultural collections.
2. Popularize fine cultural holdings.
3. Revitalize cultural heritage and cultural development.
4. Invigorate cultural, content, and value-added industries.
5. Promote knowledge and information sharing.
6. Enhance research, education, and life-long learning.
7. Improve literacy, creativity and quality of life.
8. Embrace collaboration and international community.

There are nine institutions who are participating in the NDAP, including: Academia Historica, Academia Sinica, National Central Library, National Museum of History, National Museum of Natural Science, National Palace Museum, National Taiwan University, Taiwan Historica, and Taiwan Provincial Consultative Council. More organizations, public or private, are anticipated to join the effort to contribute to this programme in the future. The NDAP budget for the first five years is $339, $368, $530, $638 and $645 million NTD, respectively, (approximately $10.3, $11.2, $16.2, 19.5 and 19.7 million USD). The second phase of NDAP will start in 2007 and covers the period 2007 to 2011. It is estimated that the budget for the second phase will be higher than the first phase.

ORGANIZATION AND FUNCTIONS OF THE NATIONAL DIGITAL ARCHIVES PROGRAMME

Programme Office

There are five divisions in the Programme Office: 1. Content Development Division, 2. Research and Development of Technology Division, 3. Applications & Services Division, 4. Training & Promotion Division, and 5. Operations Management Division. These divisions will jointly carry out the overall planning organized by the Programme Office, build common infrastructures, and establish general content processing guidelines and specifications for all participating institutions and non-solicited projects.

Content Development Division

The primary tasks of the Content Development Division are to:

1. Survey the quantity of specific types of objects collected in participating institutions, monitor the progress of digitization for each participant, and document the working procedure for each project;
2. Organize the NDAP Content Development Committee for supervising the progress of this division and working groups for contents such as paintings, rubbings, works of art, botanical specimens, etc.;
3. Build home pages for the Content Development Division and each thematic group;
4. Assist (on a consulting basis) all the participating institutions to implement their digitization projects smoothly.

The Content Development Division also plays a major role in setting up a content development strategy for future projects, establishing working standards for potential participants, and working out a mechanism for the division of labour for participating institutions.

The Content Development Division invites representatives who participate in the NDAP to form the OAI testbed team, to construct the union catalogue of the national digital archives with OAI-PMH, and to handle system technology.

Research & Development of Technology Division

The main purpose of the Research & Development of Technology Division is to carry out the plans of the programme office and to pursue research and development efforts for many aspects of digital archives technologies. This includes the:

1. Establishment of standardized information reference guidelines for digital content creation, storage, and processing;
2. Development of common and application-specific information processing infrastructure and tools;
3. Construction of the public information system;
4. Services for digital archives storage and backup; and
5. Assistance in technology skills and personnel development training for members in participating organizations.

Applications & Services Division

The Applications & Services Division, which serves as a bridge between the content holders and digital content industries, is designated to fulfil the utilization goal of sustaining digital preservation activities, promoting digital content industries and improving creativity and productivity. This division has four major missions:

1. Fund seed projects to build up experience and skills on creative-use;
2. Lay out marketing strategies to promote digital content industries;
3. Integrate resources and uses effectively to increase business transactions;
4. Promote a virtual industrial park to create agglomerated economy.

Training & Promotion Division

The Training & Promotion Division organizes training courses, conferences, and workshops for participants of the NDAP projects as well as for librarians, archivists, and people from industry. In order to promote the use of the end products of NDAP, the Training & Promotion Division conducts several workshops for elementary and high school teachers. An instructional design contest is held each year, and the contest winners are invited to demonstrate their works.

The division also attends the Taipei International Book Exhibition and conducts promotional activities with the media. Objectives of those activities are to:

1. Offer training courses for people involved in digital libraries or digital archives;
2. Assist professionals in libraries, museums, and archives to understand the development of digital archive technology;
3. Strengthen the abilities of elementary and high school teachers to use digital archive resources;
4. Invoke public interest in digital archives;
5. Promote the end products of the National Digital Archives Programme.

Operations Management Division

The Operations Management Division assists the NDAP Programme Office in various administration matters. The major functions of this division are to:

1. Provide administrative support for the Programme Office;
2. Assist external advisory committee in its evaluation of the NDAP endeavour;
3. Operate the necessary information systems;
4. Prepare various NDAP publications; and
5. Maintain a small library about archival digitization efforts in different countries.

Institutional Projects

The Institutional Projects are conducted by major content holders and providers of digital resources in Taiwan. There were nine participating institutions, including Academia Historica, Academia Sinica, National Central Library, National Museum of

History, National Museum of National Science, National Palace Museum, National Taiwan University, Taiwan Historica, and Taiwan Provincial Consultative Council.

These institutions hold national cultural heritage and collections, and take leadership positions in different areas. Digitizing their archives not only preserves cultural treasures, but it also facilitates new applications to be developed for these collections and cultural holdings.

Non-solicited Projects

Non-solicited Projects are those projects selected from proposals submitted to open Request-for-Proposal (RFP) announcements. From 2002 to 2006, 282 projects were selected, including: 90 for digital content development, 51 for technology research and development, 86 for various applications of digital archives, and 55 for creative learning.

Year	2002	2003	2004	2005	2006	Total
Digital Content Development Non-Solicited Proposals	6	8	22	25	29	90
Technology Research and Development Non-solicited Projects	5	6	10	17	13	51
Applications of Digital Archives Non-solicited Projects	16	15	25	16	14	86
Creative Learning Non-solicited Projects	N/A	N/A	10	25	20	55
Total	27	29	67	83	76	282

Content Thematic Groups & Technology Working Groups

In order for the NDAP to progress efficiently, the Programme Office has provided a common mechanism for coordination, technological support and personnel training. In the coordination aspect, 16 thematic groups and 6 working groups were formed in 2005. The thematic groups provide a coordination mechanism to solve problems arising in digitizing and managing collections in the respective areas. The themes supported by these 16 groups are: Anthropology, Archaeology, Architecture, Archives, Artifacts, Botany, Calligraphy & Painting, Chinese Classics Full-text Database, Geology, Journalism & Mass Media, Linguistics, Maps & Remote images, Rare Books, Stone & Bronze Rubbings, Video, and Zoology.

The six working groups for technology are: Reference Platforms for Digital Archives, Naming and Distributed Searching, Formats of Digital Objects and Archives, Digital Archives Services, Multimedia Digitization Process and Multilingual Information Processing. These working groups meet on a regular basis to discuss technology-related issues associated with building human-centred, scalable, interoperable, and cross-culture systems. These discussions enable specialists in the social sciences and humanities to create, access and use digital contents, and enable general users to browse, learn, and enrich their intellectual content, and hence improve their quality of life.

Outcomes of the National Digital Archives Programme

The end products of the NDAP will be collected into the "Taiwan Digital Archives (TDA)." The digital resources in the TDA are divided into three levels: the archive level, the open-market level and the public information system level. The resources of the archive level are for preservation purposes and are accessible only on a case-by-case basis. The resources of the open-market level are to be released (subject to some costs) for commercial use, from which value-added digital content and digital services can be developed. The resources of the public information system level are for educational purposes and are free of charge for public consumption. Figure illustrates a scenario of what the NDAP considers the derived benefits of the TDA to be.

In the future, besides preserving the national collections, promoting Taiwan's cultural holdings, and encouraging information/knowledge sharing among the organizations holding content and citizens, the NDAP will continue to promote new applications and services over content collections and cultural holdings. By using digital resources of TDA, the NDAP can then achieve the goal of developing a content rich and creative society, promoting industry and economy that boost creativity, productivity and quality of life in Taiwan.

Research Related Issues

The different aspects of digital library research have been considered to be very important for supporting the functions of both of the national-level digital library projects – the DMP and the NDAP. The DMP was composed of two types of project: topic-based projects and technology support projects. The major research work was carried out by technology support projects that covered different areas, including: geographic information systems, word net, metadata, copyright management, digital watermarking, and digital library evaluation. Among these areas, the author participated in and led a metadata research team, ROSS (Resources Organization and Searching Specification), to study metadata-related issues and to design metadata suitable for Chinese materials. The ROSS team also developed an XML/metadata management system known as Metalogy.

The NDAP supports research projects on many aspects of digital libraries, and these projects are carried out mainly by the Research & Development of Technology Division. This division has provided technical services for over 50 content projects, including services for metadata analysis, standards development, database development, systems analysis, systems development, and follow-up maintenance; has enabled successful technology transfers to local companies and government organizations, including a new Chinese word segmentation method to the Institute of Information Industry, a missing Chinese characters input and presentation technology to the Electronic Data Processing Centre, Executive Yuan, and the video archive construction technology to Acer Inc.; and has completed over 40 technical papers on digital archives infrastructure, information retrieval, and multimedia content management.

In the past three years, the research and development efforts have focused on intellectual property, global resources discovery, interoperability, metadata, multilingual information retrieval, communication networking, multimedia technology, information security authentication and integrated services.

The NDAP also supports some application development research projects that are being carried out mainly by the Applications & Services Division. These projects address commercialization-related issues. This division has interviewed several key commercial companies within the digital archives-related industry in Taiwan to understand the difficulties and challenges they face and to find solutions for them. In early 2005, the Applications & Services Division proposed a virtual industrial park to promote and create agglomerated economy of digital archives. This virtual industrial park provides a hub for transactions involving the products of digital archives and for licensing digital objects for commercial use.

Future Challenges

With the rapid advancement of information technologies and the increasing number of Internet users, an increasing number of museums, galleries, libraries, archives, universities, research institutes, and other organizational content holders are and will be involved in the NDAP.

As cultural digital archives have been growing dramatically in size and variety, an industry built around the construction and application of digital archives is gradually taking shape. It is therefore very important for the NADP to identify business models for the cultural digital archives industry.

From the discussions with approximately 150 companies that participated in the various national digitization programmes in Taiwan, the Applications & Services Division has summarized five categories of business activities.

They are (1) digital archives data bases, usually in Web-accessible form, (2) integrated data bases, usually on a specific subject such as paintings or plants, (3) products from creative reproduction of digital objects, (4) digital content services delivered through hardware devices, and (5) digitization technologies, software, and services. To be successful in developing the digital archives industry in Taiwan, the challenges will be:

- Establishing the environment and regulations suitable for the rapid growth of the industry,
- Attracting investment and international cooperation,
- Assisting in the implementation of investment and tax incentives,
- Enlarging the human resource pool of the innovative and managerial talent needed by the industry,
- Accelerating the development of key technologies and products,

- Assisting companies to enhance their international marketing capability and competitiveness, and
- Implementing broad applications of digital content in Taiwan.

G-PORTAL – A CROSS DISCIPLINARY DIGITAL LIBRARY RESEARCH PROGRAMME FROM SINGAPORE

The ways that average and expert users find information, and study and disseminate it, have changed significantly in the last decade due to the pervasive use of the World Wide Web. Today, instead of sending information via email or snail mail, for easy dissemination information resources are frequently posted on the Web in various forms, including web pages, document files, spreadsheets, and even blogs.

This has certainly made the Web the largest source of information, and probably the most popular one as well, assisted by easy-to-use web search engines and browsers. From the perspective of education, the Web represents an exciting medium for supporting learning activities both within and outside classrooms.

On the other hand, educators realise that many web sites contain *learning objects* of high educational value (Sumner and Marlino, 2004). Learning objects here refer to carefully edited and organized text, HTML, and other multimedia objects that can be used directly or indirectly in the learning process. For example, web pages and image files from a web site documenting environmental issues related to water conservation can be considered as learning objects.

Once created, learning objects can be made available on the Web. Web sites can also be designed such that educational software tools can be activated to support a variety of learning tasks. These learning tasks can range from having closely guided instructions to being completely open-ended. As the Web also serves as a global network of servers for user communities to create and share learning objects, these objects may be organized in some manner to allow users to take some control of these learning objects, choosing ways of organising and using them, creating new ones and even developing their own learning strategies.

Using the Web as a learning platform poses a few challenges to researchers from the education, computer and information science disciplines. While education is a highly established discipline, research on learning effectiveness within the Web environment is still in the nascent stage. Computer and information scientists have contributed significantly to web browsing and searching software, but they have few clues about the software functions needed to enhance learning, especially learning in specialized subject domains. The G-Portal digital library research programme, which began in 2001, aims to address the above challenges by developing a digital library portal for harnessing distributed learning objects and provide the essential tools to support active learning (Liu et al., 2003) (Liu et al., 2004). The research programme focuses on the geography domain, which requires content rich media including maps, images, text and other data collected from geographical environment. The G-Portal research team consists of members from the education, computer and information science disciplines, and G-Portal research activities can be divided into two complementary tracks:

a. Design and implementation of the G-Portal digital library portal with learning support functions; and
b. Deployment of G-Portal digital library portal in geography education.

In the following sections of this article, we report the work done in these two tracks, highlighting the research issues involved and the approaches taken to address them. Long-term research directions will also be described.

Design and Implementation of G-Portal

The G-Portal web portal design can be divided into database and software aspects. The former supports flexible representation and storage of metadata and their organisation, while the latter provides a set of functionalities that allow heterogeneous web learning objects to be utilized in learning activities.

Metadata for Heterogeneous Geography Learning Objects

Unlike proprietary learning objects, learning objects on the Web are created by different users for different purposes, including those that are non-educational. In G-Portal, metadata records are constructed for heterogeneous geography objects that allow them to be organised for various learning tasks. A G-Portal metadata record consists of a set of mandatory metadata elements required for uniform metadata identification, visualization and display, and optional metadata elements for storing task-and topic-specific description of learning objects.

For easy metadata sharing and manipulation, metadata schemas can be defined with appropriate mandatory and optional elements, and every metadata record is created using a metadata schema as the template. Every metadata schema must provide a mandatory source metadata element to allow metadata records to be linked to their corresponding learning objects. A G-Portal metadata record defined for a web page that contains a photo image of a beach headland. Metadata records can be associated with spatial geometrical shapes for them to be displayed on the G-Portal map interface.

Metadata Organised by Learning Activities

G-Portal allows a set of metadata records to be organized according to how they are to be used in learning tasks and by the topic to which they belong. These are known as *organization by task* and *organisation by topic,* respectively.

A *project* in G-Portal refers to a collection of metadata records constructed to support a learning activity. For example, Figure depicts a project (called "ECP Project"), which consists of a collection of metadata records about the physical structures and beach water level measurements (also called profiles) taken at the East Coast Beach of Singapore. The project has been used in a geography course to evaluate the extent and cause of beach erosion.

The set of metadata records belonging to a project can be determined when the project is created, or can be added after project creation. Within a project, metadata records can be further organized into *layers,* each of which represents a subset of metadata records that are displayed as a group. For example, Figure depicts a set of beach profiles displayed under the "ECP Profiles" layer, while the physical beach structures are grouped under a different layer.

Metadata records within a project are also organized by topic or category to facilitate metadata-browsing by category. Parent-child relationships among categories can be defined to form a category hierarchy. Unlike many other digital libraries, G-Portal can accommodate different category hierarchies for different learning projects so as to support customizable metadata organization. A metadata record can also be shared among different projects and appear in different categories. This gives much flexibility that cannot be found in traditional digital libraries.

Creating, Querying and Visualizing Metadata

In G-Portal, multiple users can access a project simultaneously and can collaboratively contribute metadata records to the project. A web form interface is available for these users to add new metadata records for relevant web objects. The created metadata can be accessed online via either queries or browsing using the G-Portal client interface. The client programme is a Java applet that runs on almost any PC or workstation. It has both map-based and classification-based interfaces for browsing metadata. Within the map-based interface, one can perform zooming and panning operations to locate desired metadata records. The client also has a query interface that can handle queries on spatial locations and metadata elements. Once a metadata record is found by query or browsing, its metadata elements can be displayed in either HTML or XML form. In either case, one can follow the source link in the metadata record to view the referenced learning object.

Deployment of G-Portal in Geography Education

Using G-Portal in a learning activity involves the following steps:

1. Construct a project with an initial set of learning objects and their metadata;
2. Access and view learning objects within the constructed project;
3. Collaboratively create and share new learning objects and/ or their corresponding metadata;
4. Analyse project content and use them in assignment report preparation.

Next we describe how the above steps were performed when G-Portal was deployed in a nature reserve field study.

The primary objective of the field study is to evaluate the environmental stress of Bukit Timah Nature Reserve (BTNR) by taking readings (*e.g.*, humidity level, temperature, root mass, etc.) and photographs at various locations in the reserve and analyzing them. The study was conducted as part of a course conducted at the National Institute of Education. Firstly, from the scope and objectives of the field study, a new project was created in G-Portal by the professor responsible for the course. The map area to be covered by the study was also determined, and its corresponding background map image was added as the background of the project's map-based interface.

In the study, a map showing the trails of BTNR was used as the background image. The stations at which data were to be collected were also indicated on the map. Students were required to form teams, and each team was to collect environmental data (*e.g.*, temperature, humidity, etc.) at designated stations in BTNR. The students were then required to create spreadsheet files and photo images about the data collected at these stations, and construct the respective metadata records. A screenshot of G-Portal's client with this project loaded, as well as its contributed metadata records.

When the data collection and metadata creation phases were completed, G-Portal was used by the students to query and visualize all the created metadata records in the BTNR project. They could access G-Portal both in the classroom and at home. Since the student teams were assigned different sets of locations (with possibly some overlaps), each student team had to use G-Portal to gain access to the metadata records and web objects created by other teams.

With G-Portal serving as the central web repository of metadata records, it was found that students were able to locate metadata records and their corresponding web objects required for the field study assignment. G-Portal also facilitated collaboration among students, as they could refer to one another's metadata records. This was also reflected in the final study reports prepared by the student teams as they cited web objects created by other teams.

In the above example scenario, all student teams shared the same project to which they contributed metadata. G-Portal can also be used in other learning scenarios where students are expected to create and update their personalized projects. To begin with, a public accessible project may be first created by the course professor to maintain a basic set of metadata records and maps. The G-Portal client allows students to then copy the relevant information from the public project and paste them in their personalized projects. The same operation can be performed on different public projects to reduce the efforts of gathering metadata records for a newly constructed personalized project.

Other than the above-mentioned online Java applet client interface, G-Portal also provides a Java application client interface that can be used when users are offline. This is particular useful (a) when users have no access to Internet, or (b) when users would like to let others access their projects without involving their user ids and passwords. This application client interface is very similar to the applet version except that this interface can operate on a standalone PC without an Internet connection.

Future Research Directions

The G-Portal research programme has so far realised a web portal for managing and sharing geography-related metadata records. Sections 2 and 3 above summarized the design of this portal and its deployment in learning activities. As part of our future research, we are extending G-Portal with additional features to support other important steps in learning. These include: (a) reviewing and rating of metadata; (b) interoperability of metadata; (c) harvesting of learning objects and construction of their metadata.

Metadata Review and Revision

Compared to raw learning objects, metadata are of better quality, as they are expected to be created by highly qualified human experts. In our proposed meta-portal system, students and community users can create their raw learning objects as part of their learning activities and submit their metadata for these objects to the meta-portal for further sharing. To ensure the metadata quality and to promote cooperation among community members, an open review system for Geography community experts will be required (Ismail et al., 2003). In designing the metadata creation and review process, the collaborative metadata editing and versioning functions will be investigated. The former allows multiple users to jointly construct different elements of metadata records. The latter supports changes to metadata records by maintaining different versions of them.

Interoperability of Metadata

Metadata is a valuable resource for learning. To fully tap into the potential of G-Portal, the metadata resources of G-Portal should be further exchanged with other l the Open Archives Initiative (OAI) protocol (Lagoze et al., 2002) can be extended with flexible metadata formats to support interoperability between G-Portal and other systems. OAI has been implemented for several public domain digital libraries including DLESE and other NSDL (National Science Digital Library) projects (http://www.nsdl.org). Because projects and layers are unique features in G-Portal, exchanging these pieces of information via OAI will also be a topic for research.

Harvesting of Learning Objects and Metadata Construction

The current G-Portal digital library largely relies on users to manually identify useful learning objects and create metadata records for them. Although the metadata created are highly relevant to the learning activities, the amount of effort required can be significant. Users are expected to be competent in finding and sometimes creating relevant learning objects themselves and competent in creating metadata records. To address these limitations, future G-Portal research will explore techniques and tools for automatically discovering learning objects and constructing their metadata from the web.

For example, we are now investigating the problem of extracting place names from web pages and associating appropriate place names with web pages (Zong et al., 2005). Our future work will focus on uncovering those metadata that are available from public domain web repositories of geography learning objects, including those provided by other digital libraries, *e.g.*, DLESE.

Our research will also investigate automatic metadata construction using machine learning techniques.

Conclusions

The G-Portal research programme represents a multi-disciplinary effort to develop a geography digital library portal for hosting metadata records and providing digital library services over them. The developed software has been put to use already in learning activities such as field studies. This article has described the design and implementation of G-Portal software and its deployment in an example field study.

Other than addressing practical issues, research will also have to investigate solutions to new problems as we extend digital libraries with new learning capabilities. Some of these research problems are also highlighted in this article. Geography is a subject that spans across geopolitical boundaries. In the fast developing Asia region, we believe that the G-Portal digital library system can play an important role in gathering metadata about the geography information of the region and making the metadata available for education.

PARLIAMENT LIBRARY AND REFERENCE, RESEARCH, DOCUMENTATION AND INFORMATION SERVICE (LARRDIS)

To keep members of Parliament well-informed of the day-to-day developments in India and abroad, Parliament Library and Reference, Research, Documentation and Information Service (LARRDIS), maintains an up-to-date and well-equipped Library backed by prompt and efficient research and reference services. The LARRDIS provides research and reference material on legislative measures, as well as on a variety of other subjects coming up before the two Houses—Lok Sabha and Rajya Sabha, so as to enable members to participate effectively in the debates of their respective Houses.

Library Service

The Parliament Library is now housed in the *new* Parliament Library Building— *Sansadiya Gyanpeeth,* which was inaugurated, on 7 *May* 2002, by the then President of India Shri K.R. Narayanan.

The new Library building located in the Parliament House Complex, is a modular, ideal, intelligent, utilitarian and centrally air-conditioned building. It is fully computerized and has optic fibre based Local Area Network (LAN) with high speed Wide Area Network (WAN) connectivity to provide linkage with other State Legislatures, Foreign Parliaments and other international organisations. Apart from the Library, Reference, Research and Computer Divisions, the Bureau of Parliamentary Studies and Training (BPST) and the Parliamentary Museum and Archives (PMA) are also located in this building.

The building has very spacious Reading Room for members equipped with modern facilities and Lecture Halls. A full-fledged Conservation Laboratory/workshop for restoring old and rare documents, a Bindery Unit, an Archival Room with temperature below freezing point round the clock to preserve audio/video materials, computer tapes, microfilm rolls, etc., a Media Centre equipped with the latest telecommunication facilities, an Audio-Visual Unit, Microfilm Reader Room, an Auditorium with a sitting capacity of around 1,100 persons, a small auditorium with facilities for multimedia presentations, a Library Committee Room and Conference Room are some of the facilities provided in the new Parliament Library Building. The Members' Reference cum Circulation Counter has been set up at the Ground Floor of the Building.

Books, periodicals and newspapers, reports of Central, State and Foreign Governments, Debates of Lok Sabha and Rajya Sabha, Bills, Acts, Reports of Parliamentary Committees Statements laid on the Tables of the two Houses, Law reports, publications of the United Nations and its allied agencies, publications from foreign countries and various other publications of reference value are available for reference and use of the members at this counter. Reference books on a wide range of subjects, arranged according to the Dewey Decimal System of classification are stacked in the first floor.

A monthly Parliament Library Bulletin is brought out to keep members informed of the latest additions of books and reports in the Parliament Library. Issue of books and other publications to members is regulated by the Library Rules which have been framed on the recommendations of the Library Committee. The Library Committee is constituted every year by the Speaker of Lok Sabha to advise him on matters concerning the Library. The Committee consists of six members from Lok Sabha including the Deputy Speaker and three members from Rajya Sabha, who are nominated by the Presiding Officers of the respective Houses.

The Deputy Speaker of the Lok Sabha is the *ex-officio* Chairman of the Committee. A Sub-committee of the Library Committee consisting of 3 members from Lok Sabha and 2 members from Rajya Sabha who are members of the Library Committee is constituted by the Chairman to select new books for addition in the Parliament Library.

The Library management functions like acquisition, processing and issue and return of books have been computerised by using the software package LIBSYS. Members can have access to the bibliographical details through the PCs at the Members' Reference Counter/ Books Circulation Counter or in the Members' Reading Room.

Members' Reference Service

The Members' Reference Service organises the dissemination of factual, objective and latest information to the members of Parliament within the time stipulated by them. The facilities provided by the Members' Reference Service fall broadly under the following categories:

(i) Supply of on-the-spot information to members from published documents;

(ii) Collection and dissemination of latest information, factual data, statistics, etc. in response to members' written reference requisitions;

(iii) Preparation of reference notes on important topics;

(iv) Preparation of bibliographies on important Bills;

(v) Preparation of Background Notes, Information Bulletins, Fact Sheets and Information Folders on topical issues for use of the members;

(vi) Maintenance of Study Boxes on topical issues in anticipation of the information needs of the members; and

(vii) Updating and printing of various regular publications from time to time.

The scope of material to be collected in response to members' reference requisitions is normally limited to subjects connected with the immediate business before the two Houses of Parliament. Members are required to hand over the written requisitions at the Members' Assistance Desks located inside the Members' Reading Room in Parliament Library Building, indicating clearly and precisely the specific points on which information is desired by them. They also have the option to convey their requirements over the telephone or direct to the officers in the Reference Wing.

The information required by members is culled out from authentic sources, arranged and edited in the form of notes or tables, as the case may be, and passed on to the members concerned.

The popularity and usefulness of Members' Reference Service can be gauged from the marked increase in the number of reference requisitions received and handled during the past few years. As against 150 references handled in 1950; 425 references in 1960, 700 in 1970, 3627 in 1980, 5167 in 1990, 6508 in 2000 and 5061 during 2003 were received and attended to both during session as well as inter-session periods.

The Members' Reference Service caters to the information needs of members both in English and Hindi as per their choice.

One officer is available at each of the Members' Information Desks, to attend to the information requirements of members on-the-spot and to assist them in filling up the reference requisitions. In order to assist members in the study of literature relevant to Bills or other topics of discussion before the House, Study Boxes are also maintained in the Library during session period and members are apprised of the same through Bulletins. All books, reports and other documents relevant to a Bill or any other topic coming up before the Houses are collected and placed together in these Study Boxes so as to make the same available to members at one place. The literature placed in the Study Boxes is not issued out during the period of the discussion on the relevant Bill or topic, but is made available for study and consultation in the Library only.

The following publications of reference value are compiled, updated and revised from time to time:

(1) President's Rule in the States and Union Territories;

(2) Council of Ministers since 1947;

(3) Parliament of India: A Souvenir (it is brought out at the end of the term of each Lok Sabha);

(4) Presidential Ordinances since 1950;

(5) India: Some Facts.

Computer Centre

A Computer-based information system named PARLIS (Parliament Library Information System) was set up in December 1985 for the purpose of introducing automation in the Parliament Library. In keeping with the need for further computerization of various services of Lok Sabha Secretariat, action was initiated in December 1991 to automate the various activities of the Secretariat, *viz.* Verbatim Reporting of Parliamentary Debates, Questions, Committees, Members' Services, Pay & Accounts, Administration, Recruitment, Reception Office, Security Services and LARRDIS. Almost all the activities of the Secretariat have since been automated.

Three Computer Centres—one each in Parliament House, Parliament Library Building (PLB) and Parliament House Annexe (PHA)—have been set up to provide necessary hardware and software support to various Branches of the Secretariat. A 'Committee of Officers of Lok Sabha Secretariat on Computerisation' reviews the computer hardware requirements of various Branches/Offices of the Secretariat.

A large number of index-based database of information generated within the Parliament was initially developed by the PARLIS to make available non-partisan, objective and reliable information and authoritative data on a continuous and regular basis to the parliamentarians. However, full text of Answers to Parliamentary Questions received from Govt. of India and verbatim Reporting of the proceedings of the House were also placed on the Parliament of India homepage. The data currently available for on-line retrieval relate to:—

- Parliamentary Questions (Full texts of Questions and Answers since February 2000; indexes from 1985 to 2000 are also available)
- In any comparison between the online catalogue and the manual catalogue, it seems appropriate to choose the cardParliamentary Proceedings other than Questions (Full text of Floor Version since the Winter Session of 1993; Indexes from 1985 to 1993 are also available)
- In any comparison between the online catalogue and the manual catalogue, it seems appropriate to choose the cardGovernment and Private Members' Bills from 1985 onwards (only indexes);
- In any comparison between the online catalogue and the manual catalogue, it seems appropriate to choose the cardDirections, Decisions and Observations from the Chair, from 1952 onwards;
- President's Rule in the States and Union Territories, from 1951.
- Members of Council of Ministers from 1947 onwards;
- Obituary References made in the Houses since Provisional Parliament.
- Library Management Functions such as Acquisition, Processing and Issue and Return of books have also been computerized using the software package named "LIBSYS". A web-based Library catalogue can also be accessed through Internet.
- Documentation Service (from 1989 onwards): (Important articles published in books, reports, periodicals and newspapers are indexed and annotated and can be accessed through Internet).

Information Available on Internet/Parliament of India Homepage

The Indian Parliament Home Page was inaugurated by the President of India on 15 March, 1996. The information available on the PARLIS database has been converted into a Web enabled environment. The major components of the parliamentary information, now available on the Internet are:

I. Business of the House (Lok Sabha)* • List of Business, • Bulletin Part-I, • Bulletin Part-II, • Synopsis of Debates, • Daily Debates, • Wit and Humour, poetry and couplets, • Questions/Answers (February 2000 onwards), • Debates (Proceedings) (July 1991 onwards)• Resume of Work, • Legislative Business.

II. Members of Parliament (Lok Sabha)* • Bio-data of members of the present House, • Information about members of previous Lok Sabha, • Members' homepages which contains information about Members' Biodata, with addresses and telephone Nos., • Parliamentary Questions asked by him/her, • His/her participation in the proceedings of the House, • Funds sanctioned for expenditure, • Incurred by him/her under various schemes such as MPLADS, etc., • Statistical Information about members, • E-Mail addresses of members of present Lok Sabha, • Addresses & telephone Nos. and date of birth of members of present Lok Sabha, • Hand book for Members, • Members of Parliament Local Area Development Scheme (MPLADS).

III. Information about Parliamentary Committees and their Reports*

IV. Parliament Library Catalogue.

V. General Information • Introduction to Parliament, • Rules of Procedure and Conduct of Business in Lok Sabha, • Directions by the Speaker, • Decisions from the Chair, • Parliamentary Documentation, • Lok Sabha Secretariat—Organization. The floor version of the text of Debates in bilingual form is available through the Parliament of India web-page. Members can retrieve the information regarding their participation as well as discussions held in Lok Sabha and Rajya Sabha on any subject of their interest. Arrangements have also been made to web-cast, the live audio proceedings of both the Houses of the Parliament, Constituent Assembly Debates, text of the Constitution of India, Information about the functioning of the Parliament and the Lok Sabha Secretariat, facilities provided to the members, important speeches of President of India and texts of Union Budgets are also available on net. The Homepage also contains the profile of the Speaker, Lok Sabha, his political and personal achievements, important speeches and photogallery, etc. The links to the websites of Government of India and to other Parliaments have also been provided at the home page. World wide access to this site can be had at the Internet address: *http://parliamentofindia.nic.in*

Computer Facilities for Members of Parliament

Keeping in view the information requirements of the parliamentarians and to automate their offices, a Laptop or a Desk top computer with a printer and a scanner has been provided to them at their residences/work places. A hand held computer has also been provided to them. Members of Parliament have also been provided Internet and E-mail connectivity through NICNET so as to provide access to the Parliament database, other Government database, database of foreign Legislatures and the vast sources of other information available on the Internet. Arrangements have also been made to provide Computer training to Members of Parliament and their Personal Assistants.

Speaker's Homepage

Speaker's Homepage, titled "The Office of the Lok Sabha Speaker" was released by the Speaker, Lok Sabha on 25 January 2002. The Homepage contains the profile of the Speaker, Lok Sabha, his political and personal achievements, important rulings/ decisions, important speeches.

Information pertaining to the development of his constituency, tours undertaken/proposed and photo gallery. Profiles of all the former Speakers are also available in this section. A link to the Speaker's Homepage has also been provided in the Parliament of India Homepage.

Video Conferencing: The Video Conferencing facility has been provided at the offices and residences of Speaker, Deputy Speaker and Secretary-General, Lok Sabha, and the Offices of Additional Secretaries and Joint Secretaries and also the Computer Centre, Parliament Library Building.

Touch Screen Information Kiosks: Twelve Touch Screen Information Kiosks have been installed at the various vantage points in PH and PHA.

Digital Library: A Digital Library has been set up in the Computer Centre to cater to the needs of members of Parliament and Officers and Staff of Lok Sabha Secretariat. A member is entitled to borrow two DVDs/CDs at a time for his own reference and use.

Microfilming Service

Micrographics is an indispensable part of information technology for better management of Library and Information Services. In order to ensure optimum utilization of the space in the Parliament Library and for better preservation and future use of valuable collections of the Library, a Microfilming Unit was set up in 1987 as one of the important Branches of LARRDIS with the acquisition of latest available models of microfilming equipments and ancillaries. The Unit is now fully functional and provides facilities for Computer-Assisted Retrieval (CAR) of information from the microfilmed documents. At present, microfilms are prepared with the help of three RV–3 and one MRD-2 Cameras having provision for 16 mm and 35 mm/16 mm width of rolls, respectively. The RV-3 Camera has the capacity to take nearly 2400 exposures from standard 16mm rolls. For operation of these Cameras, eight dark cabins have specially been designed where processing of microfilming is undertaken.

All documents are microfilmed on 16 mm rolls, except those of abnormal size, like Papers Laid on the Table of the House, which are being microfilmed on 35 mm film rolls. The Unit also has one Processor and one Duplicator for making additional copies of the microfilm rolls. The original microfilm rolls as well as duplicate rolls which are used with the help of Microfilm Reader-cum-Printer are kept in the unit. The microfilms of the documents are easily accessible to members of Parliament and to all those interested therein.

Reprography Service

The Reprography Service set up in 1975 caters to the urgent official needs of Members of Parliament, Officers and Branches of Lok Sabha Secretariat and media persons accredited to Lok Sabha Press Gallery for photocopying important Press Clippings, information contained in answers to Parliamentary questions and debates, articles in periodicals, newspapers and extracts from books and other documents.

The Service also provides facilities for Xeroxing and typing of personal correspondence/documents of members of Parliament against nominal payment. MPs and former MPs can get their documents photocopied on payment of Re. 1.00 per copy and English and Hindi typing work is done at the rate of Rs. 6 per page for single space and Rs. 5 for double space. Additional copies of the typed documents are charged @Rs. 1/-per page, from the service located in Room No. 45-A in Parliament House. The Reprography service located in G-037, Parliament Library Building also provides Xerox facilities on nominal payment basis.

Documentation Service

Documentation Service, set up in 1975 is mainly responsible for locating, collecting, classifying and indexing all the useful literature and articles in various documents, including Newspapers, Books, Reports, Periodicals, Parliamentary Debates etc. that may be of interest to the members of Parliament. All the titles are suitably annotated and arranged under a specially devised subject-wise classification scheme. This information is provided in a fortnightly publication called PARLIAMENTARY DOCUMENTATION which is placed on Internet since June 1998 and can be retrieved under different parameters like year, subject and country/state, author, document apart from specific issues/ period. The whole process of *Parliamentary Documentation* has been made online since June 2003, to facilitate the instant availability of and up-to-date information to the members of Parliament, officers and other users. The Printing of *Parliamentary Documentation* has been ceased *w.e.f.* November 2003. Since then a few computer generated copies are placed in the library for reference by Hon'ble members and other users. For the years 1975-1998, bound volumes of *Parliamentary Documentation* (earlier known as Documentation Fortnightly) are available for consultation in Parliament Library.

Press Clipping Service

The Press Clipping Service set up in 1956 caters to the information requirements of the members of Parliament on topical issues. It collects all important, relevant and up-to-date news items, editorial comments and articles on developments in the legislative, political, economic, socio-cultural, scientific and technological fields. At present, the clippings are taken from 18 English and 9 Hindi newspapers published in various parts of the country. (Besides, Press Clipping Services in various regional languages are being planned in order to enrich the press clipping collection). The Press clippings are maintained chronologically in subject folders and stacked in proper sequence according to the Dewey Decimal Classification scheme. These folders are not issued out and are consulted in the reading rooms of Parliament Library. Ordinarily, Press clippings are retained for two years. However, important clippings of lasting value and interest and those having a bearing on constitutional, parliamentary and legal subjects are, retained permanently after scrutiny. In line with the on-going process of computerisation of various activities of Parliament Library, the Press Clipping Service has started feeding into the computer all important selected English Press Clippings using the imaging technology system.

The computerised Press clippings can be retrieved through classification number and subject keywords from the Computer terminals installed in the Press Clipping Section, Members' Reference Service, Library Counter of Parliament Library Building, concerned senior officers of LARRDIS and Question Branch in Parliament House Annexe.

Research and Information Service

The Service endeavours to keep members informed on a continuing basis about the current national and international issues in various fields by regular publication (in English and Hindi) of Books, Brochures, Background Notes, Information Bulletins, Fact Sheets, Monographs, Current Information Digests, etc. Two Monograph Series known as the Eminent Parliamentarians Monograph Series and the Distinguished Parliamentarians Felicitation Monograph Series are also brought out to commemorate the services rendered by the eminent parliamentarians of the past and to felicitate the distinguished parliamentarians of today. Profiles of leaders whose statues and portraits are unveiled in the Parliament Complex are also prepared by the service.

Periodicals in English

1. The Journal of Parliamentary Information (Quarterly);
2. Digest of Central Acts (Quarterly);
3. Digest of Legislative and Constitutional Cases (Quarterly);
4. IPG Newsletter (Quarterly).
5. CPA Asia Region News Letter (Quarterly)

Periodicals in Hindi (brought out by the Raj Bhasha Prabhag):

1. Sansadiya Patrika (Quarterly);
2. Kendriya Adhiniyam Sar (Quarterly).

Copies of all these periodicals and other publications brought out from time to time are supplied to members who specifically ask for them. Copies are also made available in the Library for other interested members. The Service also prepares draft speeches/points for Speeches, Briefs, Research Notes, Resolutions, Explanatory Memoranda, etc. for Indian Parliamentary Delegations participating in various international Parliamentary Conferences held under the auspices of the Inter-Parliamentary Union, the Commonwealth Parliamentary Association, the Association of SAARC Speakers and Parliamentarians, etc. from time to time. Country Notes are also prepared for the use of members of Parliamentary Delegations going abroad on goodwill visits. Background Papers on subjects of parliamentary interest are prepared for the use of members participating in the Seminars and Symposia organised from time to time by the Indian Parliamentary Group (IPG) and the Bureau of Parliamentary Studies and Training (BPST). In short, the Research and Information Service provides the desired information on the subjects/aspects which warrant intensive and extensive research.

5

Scope in Library Profession

Daunted by the vastness and variety of the ASIA listed in the World Geographical Encyclopaedia and hampered by the lack of information on the library profession in the component countries, this paper will not provide details of any one country unless they are relevant to the issues raised in the discussion of the topic.

The focus of this paper is the library profession, particularly the enhancement of its image and status. The topic will be discussed in the following manner: (i) Scope of paper (ii) The Asian Background (iii) Defining the Library Profession (iv) Enhancement of the Library Profession (v) The Library Profession in the Virtual Library environment.

The Asian Background

In many countries in Asia, libraries have existed for thousands of years although very little has been written about them. Even in India "a nation known for its ancient and medieval, as well as more modern library establishment"-accounts of libraries in the Vedic, Buddhist, Medieval and Muslim periods of Indian history have yet to be accomplished.

In China, the earliest libraries existed in the form of "an aggregate of documents" or book collections in the keep of royal families, temples and scholarly institutions. In Southeast Asia, libraries in most of the countries are the product of the 20th century, with the exception of Philippines where the National Library owes its roots to "the fewer than 100 volumes of books gathered in the Museo Biblioteca de Filipinas established on 12 August 1889".

But modern library movement in Asia really took root in the 1970's with greater impetus in the 1980's, followed by a growth in professional literature.

A survey of the literature reveals great diversity in the library development among Asian countries, reflecting the vastness and variety of Asia itself. However, underlying this diversity are some common features that characterise library development in Asian countries:

- Inequitable distribution of libraries. Libraries are not well distributed throughout the country, with some parts being more well endowed than others.
- The earliest libraries are scholarly libraries or religious collections.
- Most of the modern libraries are dependant on Government resources.
- University and special libraries are more developed and privileged than public libraries.
- Libraries operate under decentralised management systems.
- There is often lack of cooperation, coordination among libraries, giving rise to duplication of materials as well as incompatibility of operating systems.
- Library development is spurred by economic growth, especially in cases of agricultural countries that need to industrialise quickly.
- Tendency to liberalise library use whereby closed access collections are subsequently open to public.
- Libraries become direct victims of wars or political upheavals in countries that suffered such tragedies.
- Library development is seen in connection with other aspects of national development social, cultural, economic, etc.
- Greater awareness of importance of libraries among governments in recent years.
- There is access to library education, with some more established than others.

Although as a whole, library development in Asia is impressive, enhancing the image and status of the library profession has been a matter of concern to most librarians over the last two decades. It has been discussed and written about in seminars and conferences in most parts of Asia since the 1970's. And so it should be — because the process of professionalisation should be continuous. Professional literature on professionalism has shown that no profession can be completely professionalised it can only be more professionalised as the profession treads along the path of professionalism. The process of professionalisation can also be applied to various aspects of the profession at different levels and pace so that at any one time certain aspects of the profession can be more professionalised than others.

The need to enhance the library profession has assumed greater urgency in the 1980's because of:

- greater competition posed by the proliferation of information services that are not library based.
- acceleration of IT use in library functions.
- increased professional awareness of issues related to librarianship because of greater networking among librarians.
- increase of library schools resulting in better educated librarians who opt for a career in librarianship by choice.

Altogether the 1980's saw some immediate reaction, manifested in the change in nomenclature. "Librarians" became "information professionals" or "information specialists", "library science" became "information science" and "Department of Library Science" became "Department of Library & Information Science" or "Department of Information Science". The only term that has not suffered a name change seems to be "library profession" as we witness today. Seen from a positive viewpoint however — these changes reflect the ability of the librarians to react to change and adapt to changes. But the enhancement of the library profession entails more than just a change in nomenclature. It entails having a deep understanding of what constitutes the library profession, what aspects of the profession should and can be enhanced and how they can be enhanced, bearing in mind that the goal of enhancement is increased professionalism. Professionalism in turn breeds excellence.

Defining the Library Profession

What exactly is the library profession? Is librariansh ip a profession? Can we call the work we do — acquisitions, cataloguing/classification, indexing, abstracting, information retrieval (with or without technology), user education, reference — professional? Is library work "nothing more than the application of sets of skills and techniques" or just "a study of systems"?

It cannot be denied that the term "profession" is difficult to define. It has been given various definitions by various people at various times. It is an elusive term and various people have attempted to define it using their occupations as a basis, resultin in definitions that are coloured by occupational bias and vested interest. When Melvil Dewey stated that "The time has at last come when a librarian may, without assumption, speak of his occupation as a profession"

little did he realise the struggle for professionalism that he had begun would continue today.

Contemporary understanding and usage of the term can be said to date from 1915 when Abraham Flexner suggested some criteria as the basis for determining whether or not social work could qualify as a profession. He suggested that a profession is:

- Intellectual and carried with it personal responsibility for the exercise of choice and judgement.
- Learned because its exercise was based on a substantial body of knowledge which could be passed on from generation to generation from practitioners to students.
- Practical in that its corpus of knowledge is put to a practical use of benefit to others.
- Organised into associations of practitioners.
- Characterised by an idealism which in theory, if not in practice, puts the aims and practice of the profession above mere money making.

Since then, Flexner was followed by several other exponents of the "traits" or "attributes" method of defining a profession, such as Carr Saunders & Wilson (1933), Morris L. Logan (1953), T. Parsons (1959), G. Millerson (1964), etc.

Together the attributes that they had put forward as worthy of a profession themselves could fill a thesis but the recurring attributes could be summarised as below:

1. Possessing a corpus of professional knowledge comprising theories and techniques/skills, preferably of a multidisciplinary nature.
2. Formal system of education and training, sufficiently long to enable the mastery of theories and techniques. This system should incorporate continuing education programmes and emphasise research and publication.
3. Possess a service principle that places the welfare of society above personal gains.
4. Be governed/regulated by a Code of Ethics, to ensure accountability in the performance of duty.
5. Maintain standards in all aspects of professional activity, such as work standards, educational standards, personal integrity, etc.
6. Be represented by a professional association.
7. Possessing legal and public recognition of professional status.

These attributes of course are not listed in order of priority and one is not more important than the other. A profession may have all or some of the attributes, depending on the level of professionalisation. The 'attributes' above correspond to a large extent to the attributes of other professions, and in general have been accepted as a popular means of assessing the level of professionalisation that a profession has undergone.

In the same manner, these attributes will serve as a basis for the discussion of the topic — enhancement of the library profession. On the assumption that these are the attributes of the library profession, how could they be used to enhance the profession? In the language of management today, these attributes can be said to be the Critical Success Factors (CSF's) of professional development.

ENHANCEMENT OF THE LIBRARY PROFESSION

Expanding and Developing the Corpus of Knowledge

Every profession needs its own body of knowledge (theoretical foundation and specialised skills) to be exclusive, such that it sets it apart from other professions and establishes its identity as a profession. Medicine, Law, Engineering, Education, etc. have their own core of knowledge and mastery of their specialised core set them apart from each other and from the para professional group within their own category.

This core makes them so exclusive that, as an example, even if the doctor were to sit under the coconut tree with his stethoscope, he can still practise his profession. Can we say the same for the librarian? Can a fresh graduate without library qualifications undertake cataloguing and classification, indexing, abstracting after being trained for a week? I say yes! All he needs is intelligence, subject knowledge, general knowledge and the ability to look up AACR II and LCSH. Now with OCLC and Bibliofile CD ROM databases, why does one need the librarian? And if the fresh graduate can do what any librarian is trained to do, what then, is so special about librarianship?

The lack of the intellectual foundation has long been the weakness of the lbrary profession. This has to be rectified if the library profession is to be enhanced. If we were to scrutinise the intellectual core of other professions, we will realise that they are generally multidisciplinary in nature and that they comprise theories and skills. In education for example, the theoretical basis is very broad encompassing psychology, sociology, management and administration, development, history, while at the same time pedagogy provides teachers with specific skills in teaching. What about librarianship? Would it not be sensible for us to broaden our intellectual base to include Psychology (to study user behaviour, educational psychology, research psychology), Sociology (to understand the cultural/social environment/ framework within which the library operates) Philosophy, Local History, Fine Arts, Communications, Languages, Law, Management, Computer Science/IT (including programming) while Research Methodology, PR, Indexing, Classification,

Information Retrieval are examples of the professional/technical skills that could be taught. The intellectual core in any profession does not merely provide facts but trains the person to reason, rationalise, solve problems — in other words, to think. Independent thinking is one dimension that, for now at least, the computer has not been able to substitute. Taking for example, the study of History, it is not the facts that are so important but the lessons learnt from the past. One learns to reason out the causes and events, understands how people think and behave at that particular moment in time and evaluate, with the wisdom of hindsight, whether or not certain events could have been avoided.

Expanding and developing the corpus of professional knowledge would place the library profession competitively with other professions. Mastery of it would provide librarians with identity, authority and autonomy — no one else could encroach into their domain. The lack of recognition that librarians in Asian countries suffer from is partly because they are constantly being compared with the more prestigious professions, such as medicine and law and engineering.

In most parts of Asia, the status of librarians suffer vis a vis doctors, lawyers, engineers. Even at universities, librarians have not been granted parity vis a vis the academic staff although India has been somewhat fortunate albeit not without a struggle. In Philippines, the status of librarians is safeguarded by legislation but these are only two out of the numerous others. In Japan, although libraries have had a long history, "librarianship has never been viewed as a professional in the European sense. Such is the case even now."

Enhancing the profession through the expansion and development of the corpus of professional knowledge is tedious but if medicine can survive the years of development, why can't librarianship? The fastest way of developing the corpus would probably be through the educational system, particularly through research and publication.

Upgrading Library Education and Training

If librarianship is to be accepted at par with other professions, library education must be undertaken at tertiary level. Advocating librarianship as a university discipline, at least in Asia, is not for snob appeal. It is a pragmatic and logical step to take because in most Asian countries, the salary scale is tagged to the qualifications obtained. Another reason why librarianship should be taught at universities is because research and publication activities are part of the lecturer's contractual obligations. Through research and publication the profession will be developed and enhanced. Yet another reason is the opportunities provided for continuing education at universities, such as postgraduate programmes (Masters, Ph.D.), seminars and conferences, study leave, etc.

University education could enhance the profession by controlling the entry qualification and providing the accreditation so crucial to maintaining standards within the profession. Library education in India, China, Japan and the ASEAN countries has developed tremendously over the last decade. Mainly conducted at universities, they undertake curricular reform to keep up with the latest developments. As example is China, where "traditional subjects [were] being deleted and new ones such as Information Theory, Library Automation and Cybernetics [are] being introduced". If developing the corpus could enhance the profession, upgrading library education and training would provide librarians with the means to propel the profession to greater heights.

Review the Service Principle

The altruistic "no profit" service ideal that has governed library services for far too long is obsolete. It will be replaced by another the "cost effective" principle. For the idealists, it will be the end of a scholarly tradition for the pragmatists, it is not too soon. In countries where libraries are heavily dependent on the Government Treasury for every cent, it would seem logical to not only find alternative funding but to introduce fee based services as one alternative. While to the purists or idealists levying charges for library services is tantamount to blasphemy, most library managers must realise that the exorbitant cost of maintaining good collections and service for free cannot be justified.

Enforcing the Code of Ethics

Most professions are regulated by the Code of Ethics but unless enforced, the Code of Ethics will not acieve its objectives.

Will the Code of Ethics enhance the library profession? If it succeeds in helping the profession to gain the respect and confidence of the public at large and its clients in particular, then the Code of Ethics will help to enhance the profession. But the difficulty with the library profession is that information work does not have the same "clout" as medicine or law. It is logical to assume that, using medicine and law as role models, accountability hinges on the protection of life and property. We do not protect life or property and who bothers about wrong information? Especially if given free! However, in the event users are charged for services/ information rendered, then library clients would presumably demand value for his money. Librarians would then have to be accountable.

Setting and Maintaining Standards

Standards are crucial in the process of professionalisation because they represent quality and excellence. Like in other professions, standards for the library profession should not be static but should be upgraded as the profession becomes more professionalised. Standards are expressed in various ways. The mission statement of any institution itself is a standard. Standards can be set for work, behaviour, services, productivity, management, education, etc. Standards breed excellence, trust and respect and eventually earn librarians public recognition and confidence.

Professionalizing the Professional Association

The role of the professional association in promoting the profession is well documented in the professional literature. It serves as the mouth piece of the profession, its representative and depending on its strength, provides leadership. However, whether on not the association can effectively undertake its role depends on several factors — level of professional awareness and support among its members, legal and public acknowledgement of its role, its political clout and most important — its ability to influence members of the profession as well as the public of its authority and ability to control matters of professional interest.

One way of assuming control and authority is to establish systems of control such as accreditation or qualifying board. In most countries in Asia however, the professional associations do not have the ability or capability to assume a leadership role and this renders the association incapable of enforcing the standards it may want to impose. Another weakness is the inability to command loyalty from its members because the institution employing librarians usually expect and do get their loyalty from their employees.

If the association is not in the position to command loyalty and assume leadership, can it be effective in the enhancement of the profession? In Asia, the strength of professional associations lies in its promotional role. They have been very successful in promoting continuing education (courses, seminars, talks, training) as well as producing publications. In some countries, despite the dynamism of the associations, they are not able to provide leadership.

The Malaysian Library Association is one such example. Dynamic though it is, registered as a society, under the purview of the Registrar of Societies, it has to function as a society unlike the Malaysian Medical Association, the Malaysian Bar Council or the Malaysian Institute of Accountants.

Attaining Legal and Social Recognition of Professional Status

Attaining legal and social recognition of its professional status would definitely enhance the library profession. However, except for Philippines, thus far the only country in Asia that has achieved legislative status, it has remained elusive for the others.

While legal recognition is clear cut, social recognition depends on how society views the contribution of librarians and library work. In this aspect a lot would depend on how librarians project their role.

Currently, librarians call themselves "information professionals". To get society to recognise this role, librarians will have to prove that they actually undertake information work. The onus is thus on the librarians to show what they are capable of. This is where librarians must use whatever tactics they have to project themselves, based on ability and quality. There is evidence of the librarians' willingness to change and adapt to the challenges posed by developments. Many have adopted new concepts from other professions and applied them to librarianship. "Strategic Planning" is one, "Performance Measurement" is another. Both have currently became part and parcel of library management.

The Library Profession in the Virtual Library Environment

The electronic library is a reality in Asia and to what extent they can be networked to form the virtual library with the rest of the world depends on will and wealth. In many of the countries, the electronic library network is already in existence. Through Internet the whole of Asia will be networked.

Would global networking however enhance the library profession? To those who believe in the power of technology they believe the future of the library profession lies with technology. To those who believe computers are just means to an end (meeting user needs) they fear that the library profession would die a natural death when libraries are replaced by professionals who, via computers, can do the librarians' work faster and more effectively. To the latter, technology is something that they fear. Ironically, doctors welcome laser surgery and computerised diagnosis/prognosis as new methods that would advance and refine their skills. But librarians view computerised indexing and cataloguing as new methods that would gradually replace their conventional skills and displace them. In a way they are correct if librarians believe that indexing and cataloguing are all that is librarianship.

In the final analysis, whether libraries are digitised or not, the image and status of the library profession depend on what librarians perceive the library profession to be. The future of the library profession rests with us librarians.

Library Information

The University Library provides facilities and services to its faculty, students, staff, and other scholars and information users through a system of discipline-and research-based libraries. Generally the University Library hold more than two million catalogued items consisting of books, periodicals, and large amount of electronic databases. In addition, they contain extensive collections of phonograph records, maps, and manuscripts. They receive a broad selection of periodicals, newspapers, and other serial titles, and sometimes they can also serve as the regional depository for that country's government publications. Based on the survey of five-university library on the Internet, the following are some of the information items to be included in the university library Web site.

- Parent Organization
- Aims and Objectives
- Maps & Directions
- Rules and Regulations /Policies
- Membership Rules
- Organization and Administration
- Library Departments
- Working Hours
- Library Directory & floorplan
- Staff & Phone Directory
- User Orientation/ Directions
- Branch Library
- Library Statistics
- Library Budget

Parent Organization

A brief information on the parent organization and the role of the university library in achieving the objectives and goals of the parent organization.

Aims and Objectives

Each and every library is established to fulfil some aims and objectives of the parent organization. Without aims and objectives no library can exist. The goals of the library are stated here. Generally the aims and objectives of the university libraries are

- To support Academic and Research Programmes
- Provide a single access point to a wide variety of on-site and networked resources in all formats.
- Facilitate access to library services for on-site and remote users.

- Provide on-line instruction to users about research processes and the effective identification and use of information resources.
- To utilize appropriate, up-to-date technology
- Take advantage of new developments as appropriate and fiscally possible.
- Support integration of capabilities with other information systems within the libraries and the University and with remote information systems.

Maps & Directions

Maps and Directions can give entire structure of the university Web site and it helps the user to reach the university safely. If the structure of a site is too complex then it guides the users to find out the particular information required by the users. If the site is not too complex then too it is better to provide site map as it helps the users to a particular page of a Web site without going through the other pages which are required by them. It will also provide the geographical location of a library, or in other words, it informs about where a library is exactly located. It also helps the users to visit the library if they are interested.

Rules and Regulation/ Policies

General rules and regulations of the university library should be stated here. Policies may include decisions regarding acquisitions, collection development, resource sharing, electronic documents, etc.

Membership Rules

Who can be members of the library. In the context of Internet there can be a variety of members. The library can have members on the Internet with limited access to the information available.

Organization and Administration

All university has its own organization and administration. A brief information should be given in the Web site. Organization and Administration includes Organization Chart and Committees and Workgroups.

Committees and Workgroups

There are several committees and workgroups within the university infrastructure. It is very important to know all those as it will help in answering some queries and according to the need patron can meet to the respective committees and workgroups. All university web sites should describe all those to avoid further confusion to the users. The structure of all the committees and workgroups are described under the following headings. Committees and Workgroups includes.

University Library Committee

It includes:

- *Self-perpetuating Committee:* The self-perpetuating committee is a creature of particular statute, which establishes the libraries themselves. Such a committee has got the sole authority and is independent as regards the control and management of library under it. It has not to report to any other higher body about its activities.
- *Ad hoc Committee:* An Ad hoc committee is a special committee for supervision and control of libraries. This committee is more or less independent and it has nothing to do with the Local Government Bodies. This committee is also called statutory authority committee. It has the advantage of being independent of politics. It takes decision expeditiously. It can co-opt men of intellect and foresight. This committee serves as the library authority.
- *Nominated/ Elected Committee:* A larger committee or an authority nominates or elects a smaller body for looking after certain bodies under it. It delegates certain functions to such a committee.
- *Recommending Committee:* This type of committee is found in England. It does not have any real power except that it simply gives certain proposals, which are subject to the approval of the library authority.
- *Reporting Committee:* This committee has sufficient powers to decide the matters within certain limits. Such decisions need no confirmation of the supreme authority but the decisions are reported to the latter for its information.

- *Executive Committee:* This committee is the most powerful of all the types except that of a self-perpetuating one since it has full powers in those matters which are delegated to it. It need not report its decisions to the library authority.

Administrative Council

It includes Budget, Planning and Assessment.

Library Budget

The import of the many definitions given in library literature bring to the fact that the budget apart from being a financial plan of the estimated revenues and expenditures for a given period of time, is forecast of the means of carrying the plan into effect, and is also an indicator of the amount of money assigned to a particular purpose.

Types of Budget

- *On What Budget:* This is also called as the item Budget/ Line item Budget. In this budgetary statement, various statement, various items of expenditure-salaries, cost of documents, cost in relation to binding, photocopying service, maintenance, etc.-are listed successively. The extent of breakdown given in regard to each item of expenditure is different in different organizations and depends on the policy or procedure followed by the concerned organizations. This budget may be an annual recurring or capital budget.
- *Formula Budget:* Here the association of resources is based upon some known or assumed relationship between two or more variables which are pertinent to the services rendered. In this type of formula budgeting one can only determine the amount the library should get and how it should be spent.
- *Unit-Cost Budget:* It is also called as Performance Budget and is based presumably on what the unit does and what quantity of services it proposes or is required to perform. Here unit costs are calculated for each item of work or activity.

- *Lump-Sum Budget:* In this type of budget, the organization provides a certain amount arbitrarily. This apportioning is not based on any logic or rhythm or reason. It depends on the whim and fancy of the management concerned.
- *Programme Budget:* In this kind of budget, the proposed expenses are categorized by functions to be undertaken. It is a "To what purpose" budget and indicates to the management how the money allocated is expensed. A sophisticated version of budgeting is called as Planning-Programming-Budgeting System (PPBS) which aims at analyzing the expenditure in relation to purpose by first defining objectives, then examining alternatives for achievement of objectives with corresponding costs before finally deciding the course of action to be taken. The purpose of PPBS, was to serve as a means of providing policy makers with an analytic evaluation of existing and proposed activities, whenever possible, with an quantitative measures of performance-that is, with costing of library operations. In other words, PPBS is a method of budgeting which aims at analyzing expenditure in relation to purpose and not by items of expenditures (salaries, documents etc.), and relates it to the results achieved. According to this method of budgeting, the components would be
- Objectives of the organization,
- Activities (and analysis of procedure) which relate to the objectives,
- Costs,
- Benefits and by-products,
- Analysis of alternatives.
- *Operating Budget:* The operating budget is a statement of estimated revenues and expenditures for a given period of time, say one year. This shows the annual recurring expenditure in relation to staff salaries, documents, and other expenditure, such as, binding, printing, etc.
- *Capital Budget:* The capital budgets are of the non-recurring type since items in this budget do not find a place, generally,

in the annual budgets. They may be considered as part of long range plans of the organization. The items included under this are the costs in relation to buildings, furniture and equipment which are considered as assets.

- *Balance-Sheet Budget:* Balance-sheet budgets are those that give information pertaining to the amount still available for use for achieving particular goals on a stated date. In other words, it can be said that these budgets serve as aids in the management of the operating and capital budgets.
- *Zero-Base Budget:* Zero-Base Budgeting (ZBB) is a method of budgetary planning and presentation which includes objective setting, programme evaluation and operational decision-making involving managers and staff. It requires that the cost of all programmes, both current and new, be justified at the beginning of each budgetary cycle. In each fresh budgetary cycle, the previous sums appropriated are 'set to zero' and each proposal is justified afresh.

Components of ZBB

- Identification and defining of decision units
- Assigning costs and benefits
- Preparation of decision packages for the decision unit
- Decision-making.

Library Assembly

It includes:

- *Book Selection Committee:* The book selection work is a very responsible job. The success or failure of a library's programme depends upon it. A Book Selection Committee in university libraries generally consists of the Heads of various teaching departments, but in practice only two or three senior staff members of an academic institution constitute the Book Selection.

The selection procedure is as follows:

- Suggestions of the Heads of departments,
- Suggestions of the members of the library committee,

- Suggestions of the library staff,
- Suggestions from the readers.
- Book Sale Committee

Under the auspices of and in close cooperation with the Friends of the Libraries, the Book Sale Committee will hold sales of donated books, recordings, and other library materials except periodicals to raise funds which will be divided between the Library Staff Association scholarship fund and the friends of the libraries endowment fund.

The activities of the Book Sale Committee is governed by guidelines approved by the Assembly after consultation with the Dean of the Libraries and the Director's Advisory Group. The guidelines must provide for the early choosing of dates for the sale that do not conflict with university-wide activities such as Honours Week or library professional organization meetings which numerous staff are expected to attend. Here minutes of the last three meetings should be given with the recommendations. The names and duration of the committee should be given properly with the contact addresses.

- *Election Committee:* The Election Committee will conduct elections as required under a certain Article and Section/s. They will design, distribute, collect, and count all ballots; retain the certified tabulation of the election results; and publish the names of the newly elected representatives. The names and duration should be given with the contact addresses.
- *Public Relations Committee:* The functions of the committee will include the following: to highlight areas of interest and concern related to the public's awareness of the Libraries; to assess and encourage support for the Libraries among campus and Metroplex communities; to determine how and when to use such resources of the Libraries; and to secure appropriate administrative approval for all activities of the committee.
- *Publications Committee:* The Publications Committee will oversee all publications produced by the university libraries, in order to avoid conflicting information and to

maintain a consistent quality of production. Individual publications will have their own editors, but the Publications Committee will act as the final editorial authority.

- *Scholarships and Awards Committee:* Each year, the Scholarships and Awards Committee will: determine how many libraries scholarships will be awarded to full-time staff and how many to part-time staff; determine the amount of each scholarship; and, at least one month before the deadline announce the opportunity to apply for the scholarships. The scholarship recipients will be chosen by the University Scholarship Committee. The Scholarships and Awards Committee will also arrange the selection of Outstanding Staff Awards recipients, as well as the Student Assistants of Distinction Awards.
- *Staff Development Committee:* The Staff Development Committee will develop training activities and programmes to help employees of the libraries perform their jobs better. These activities shall be consistent with library goals and staff needs and shall concern subject matter not covered in the technological training provided by LAN Management.
- *Review Committee:* This committee will review all the matters concerning the libraries. They are responsible to the Vice-chancellor of the university.
- *Circulation/ Reserves Work Group:* This committee will look after the circulation and reserve policy of the university library.
- *Collection Development Work Group:* This committee will look after the collection development policy of the university.
- Digitization Work Group: The functions of this group are:
- Collaborate,
- Share equipment and software,
- Prioritize projects,
- Coordinate and brainstorm future projects (archiving, core collections),

- Provide cost projections,
- Make in-house vs. outsourcing decisions,
- Maintain familiarity with all Libraries digitization projects,
- Pursue joint ventures with other departments and entities,
- Keep staff up-to-date on relevant web sites, projects at other universities, and their own projects,
- Investigate and take full advantage of funding sources,
- Determine equipment and software needs,
- Maintain an available equipment inventory.
- Electronic Information Resources Work Team

It includes:

- *Cataloguing Implementation Work Group:* The objective of this work group is to choose the actual cataloguing code to be implemented in the library.
- *WebPAC Work Group:* The objective of this work group is to help the users by their public access catalogue.
- *Library Liaison:* This group deals with all sorts of library liaison activities.
- *Reference Services Work Team:* This work team deals with giving reference service to others.
- *Web Site Development Work Team:* Mission of the Web Site Development Work Team: To provide directed access to quality information and promote the services and resources of the University Libraries.

Library Department

This section includes what are the information services and service units that cater to the users and the concerned contact persons. Library Departments are of following types:

Archives Department: The archives contain mainly manuscript collections. The archives also hold over one thousand oral history transcripts concerning various historical topics. It can be under University Records, Historical Manuscripts and Oral Histories.

Circulation Department: The circulation department checks out both circulating and reserve materials, places holds or locates on checked-out items, issues interlibrary cards, retrieves items from the Library Annex, collects fines, maintains checkout and patron records, and monitors the security, opening and closing of the building.

Government Department: The collection of Government publication is the sole objective of this department.

Humanities and Social science Department: The purpose of the Humanities and Social Sciences Department is to provide assistance with the information and research needs in the humanities, social sciences, business, and education.

Interlibrary Loan and Document Delivery Department: Interlibrary Loan supplements the University Libraries' materials by requesting items from other institutions.

LAN/ PC Management Department: The Libraries' LAN/PC Management Department is responsible for staff training, the management of the Libraries' network, personal computers, CD-ROMs, peripheral devices, and the Willis computer labs. It also coordinates activities with other campus computing entities.

Media Library: The Media Library serves the faculty, staff, and students by collecting, maintaining, and providing access to a variety of media materials.

Periodicals Department: This department deals with Periodicals section of the university, responsible for all sorts of house keeping operation for journals of the library.

Reference Department: This department deals with giving reference service.

Special collection Department: This department deals with special collection of the university in concern.

Photocopy Service Department: This department deals with photocopy service.

Rare Book Department: This department deals with managing rare book collection and helping in research activities to the researchers.

Science and Technology Department: Within each department mention should be made about the amount of collection, services, and hours of service etc.

Working Hours: The working hours of the library is mentioned here along with the holiday list. After visiting the Web site of a library if one is interested in visiting the library then it helps them to decide when to visit the library. Generally in University Library the hours of service 80-100 hours/ week.

Library Directory & Floorplan: Here a complete directory of different branches of the library with their position (*i.e.* in which floor) should be given. If possible a complete map of each floor with the indication of each department can be given for better understanding.

Staff & Phone Directory: A complete list of Staff Directory can be presented in order of hierarchy. The items should include here are Name of the Staff, Designation, Phone No., Appointment time, e-mail address etc. Under each department the names of the staff should be organized alphabetically.

User Orientation/ Directions: This is absolutely necessary and it especially should be targeted to new users. If there is online registration of users, these pages can be automatically displayed. In addition, some of the pages can be targeted to prospective users and these will be highly useful in making a potential-user a user.

Branch Library: Here a small description about the existing branch libraries within the purview of university can be given. The other information include type and amount of collection, services offer, hours of service, rules and regulation if any etc.

Library Statistics: Library Statistics will cover the information like Collections, Staffs, Expenditure Materials, Salaries and other statistics.

- Collections

It provides the collection statistics of the university library and highlights its strength. It can give information like:

- Total collections (in volumes) in each type of library,
- Volumes added,
- Monographs purchased,
- Current serial titles,
- Government documents (total),

- Microforms (total),
- Computer files (total),
- CD-ROM (total),
- Manuscripts/Archives (total, in shelf-feet),
- Maps (total),
- Audio materials (total),
- Graphics, *e.g.* photographs (total),
- Films/ videos (total).
- Staffs

It need not include the names of the staff. But inclusion may give a personal touch and the staff would be much more aware of users' impressions. However, the staff details may give an idea to the user regarding the size of the library and its various activities.

Library Budget: Provide the annual budget of the library along with the budgets of the recent past, preferably in graphical form. This may help users become aware of the dwindling budgets and may thus bring pressure on the authorities.

Contact: Contact persons and their addresses including e-mail addresses should be given. It will help in getting feedback from the user and sometimes it will help in knowing authenticity of the information.

Search Facility: If the user is looking for specific information that is not reflected in the content page of the Web site, an additional search facility will be of great help.

Annual Report: An annual report is the survey of the actual work done during the preceding year. An annual report is, perhaps, the best medium through which the Librarian can know the achievement and shortcomings of the library services alongwith the reasons thereof and thereby he can make requisite efforts to speed up the developmental activities. Annual report can attract the readers of the library. It can contain the information about the research in progress and the research undertaken. This particular information helps the researchers in avoiding the duplication of research. The annual report of a library should contain the information like:

- Clientele,
- Book Stock,
- Technical Service,
- Finance,
- Miscellaneous.

Library Services

This section provides information about different types of services provided by the library to satisfy the needs of the users. Some of the services which most of the university libraries provide are:

OPAC

Circulation Service

What's New

Current Contents

Books of Interest

Reservation Service

Documentation Service

Interlibrary Loan Service

Online Search Service

Reference Service

Referrel Service

Indexing and Abstracting Service

Current Awareness Service (CAS)

Selective Dissemination of Information (SDI)

Bibliographic Service

Webliographic Service

Publications

Supporting Service

Instruction Service

Bulletin Board Service

Special Collection Service

CD-ROM Service

Online Book Suggestion

Access to Internet and E-mail

Links to other library sites/ OPACs

Intranet Implementation

OPAC: Online Public Access Catalogue (OPAC) is the most widely provided information service on library Web pages. It is hardly the case that a library has a Web page without OPAC. It helps users to browse/ search the catalogue remotely, even from homes. It also gives an idea to the user about the status of the document whether it is issued or available.

Circulation Service: The circulation function in libraries involves every use of library materials and the successful performance of the function is a measure of the effectiveness of all other duties to which librarians devote themselves. The best-equipped library located in the most modern plant and staffed with the most capable librarians would be a mausoleum of recorded knowledge if the circulation function were omitted. The processes and services, which result in bringing users and library materials into productive relationship, are the circulation function in libraries. The circulation functions include hours and telephone number of the circulation desk, lists of eligible borrowers, Responsibilities of the borrowers, Loan periods, Charging Discharging procedures and Renewal procedures.

What's New?

Similar to the "Recent additions" or "Monthly additions", to the library, this service helps the users become aware of the recent acquisitions. One advantage of this facility is that it can be updated easily. Images of cover pages of the books can be kept, without having physically removed them from books. In addition, reviews of the books collected from various magazines can be displayed in these pages.

Current Contents: Content pages of the recently received issues of the journals can be provided, either as an image or as a text, hyper-links can be provided to the abstracts of the articles.

Books of Interest: The university library can compile a Web page giving information about the recent release of books that users would be interested and might suggest them to be acquired by the library. Such information can be collected from the publishers' catalogue or from the known Web sites of the publishers. This greatly helps the library collection to be uptodate.

Reservation Service: In case a document has been loaned, then a user, who needs it, can get it reserved. So that when the document is returned then the user can be informed and he/she can get it issued. Online reservation facility should be there as many users need not come to the library to reserve a book, in that case online reservation facility helps them a lot.

Documentation Service: The art of collecting, classifying, organizing and dissemination of information is called documentation. It consists of documentation work and documentation service. Documentation work is the process of preparing documentation list. It is provided in anticipation of demand, whereas documentation services are provided on demand only.

Inter-library Loan Service: An inter-library loan is one made by another library to the University Library, so that a researcher in the university may have access to materials which the university does not own. Copies, rather than original materials, may sometimes be supplied.

Online Search Service: The ideal university library should provide customized searching of computerized databases. Through online vendor the library can access various types of databases, covering a variety of subjects. The databases have access to journals, books, conference papers, reports, government documents and other sources. Most have a printed counterpart and the online files are usually updated more frequently. Some databases do not have a printed counterpart and online searching is their only access. Computerized searching is faster and has the ability to search for more than one concept at a time. The service is designed to meet the information needs of the university's students, faculty and staff in that order of priority. Since the library utilizes the computerized literature searching to supplement its own collection, costs will vary according to agreements made, database searched and the status of the patron.

***Reference Service*:** Reference Service is the hub of all the activities of a library. Reference service is ' a sympathetic and informed personal aid in interpreting library collection for study and research'. Samuel Rothstein gave the following distinguishing criteria of reference service. The provision by librarians of personal assistance to individual readers in pursuit of information. The recognition by the library that such assistance is an indispensable means of a definite responsibility to provide it. The existence of a specific administrative unit to furnish such assistance comprised of personnel specially equipped in the techniques of reference work.

According to Dr. Ranganathan, "Reference Service is the prepotent achievement of modern humanism in library science and it is a contact between the right reader and the right book at the right time and in the right personal way".

The objectives of Reference Services are:

- Providing of information using reference books in response to specific queries of the users.
- Giving instructions in the use of, and guidance in the choice of documents.
- Giving referral service for inquires which the information centre/ system can not answer.

According to Dr. S. R. Ranganathan there are following four kinds of reference service.

These are:

- Initiation of the freshman.
- General help to the general reader
- Short-range Reference Service.
- Long-range Reference Service.
- *Initiation of the freshman:* When a new reader becomes a member of the library he is initiated into the technicalities of the library and large stock of the library. If he is left totally unnoticed, there is a danger that he may feel

bewildered and may not be able to use the library resources. It is essential to create confidence in him so that he uses the library without doubt or fear. He should be properly guided as to the location of various kinds of books. He should be told the method of finding out the requisite reading material. Readers of this kind should be fully guided as to how to become a member of the library and to use its resources.

- *General help to the General Reader:* Readers of this kind are not new to the library but they still want some kind of help so that they may be able to get their requisite reading materials. They must be told as to how to use the catalogue and various sequences of the library. It means that such readers do not ask any particular question but they would like to know the methods of using the library resources in a better way.
- *Short-Range Reference Service:* Short-range Reference Service consists of providing readymade answers to particular questions. Time is the essence of his kind of service. No special effort is to be made to find out the requisite information.
- *Long-range Reference Service:* This kind of service is just the reverse of short-range reference service. To cull out information, lot of time has to be devoted and the sources from which information is got out are not ready reference books but lengthy reports and treaties, which need, prolonged search for finding out the adequate and proper information.

Referrel Srvice: Normally reference staff attempts to bring users in contact with documents likely to contain required information. Similarly reference staff can tell about the persons and institutions, which may be able to provide information to the users seeking information.

Indexing and Abstracting Service: Indexing and abstracting, which follow preliminary selection and acquisition of documents, involve the process of assigning descriptors to each document to identify the data or information. This would involve a comprehensive and highly condensed presentation of information.

This results in indexing and abstracting services/ products that indicate to a user the type of data/ information contained as well as parameters of data points. Some of the products emanating from these services are, Abstracts of technical papers, Patent abstracts, Digest for management, Digest for technical division, Digest for operator, technical note etc.

Current Awareness Service (CAS): Current Awareness Services are devices meant for the speedy announcement of newly acquired information or documents. Timeliness if the essence of this service, and it is therefore also called as an alerting service. The main objective of this service is to keep the research, development and management personnel of the university and other interested persons to get the current development in their respective fields of interest as quickly as possible. Current Awareness Services are to be designed with the main emphasis on speed of announcement, ease of use, and minimum time required for compilation work.

Some of the Current Awareness Services that have been developed are: Current Titles, Research-in-progress bulletin, Selective Dissemination of Information (SDI), Newspaper clipping service, etc.

Selective Dissemination of Information (SDI): H.P.Luhn is a pioneer in this field. According to him, "SDI is that service within an organization which concerns itself with channeling of new items of information from various sources to those points within the organization where they can usefully serve someone's interest. It endeavours to prevent indiscriminate distribution of new information and avert the resulting danger of not communicating at all". The basic concept behind SDI consists of matching information/ documents with the profile (interest) of each individual of the clientele. The profile can be single user or a group working on the same project or some limited subject field. Those item matches are brought to the attention of the user. In an automated system this service can be performed effectively. The aim being that user should neither be provided too much information nor made to miss information essential towards his requirements. The steps involved here are:

- Users Profile,
- Documents Profile,

- Matching,
- Notification,
- Response or Feedback,
- Readjustment or rejuvenation of Profiles.

Bibliographic Service: A bibliography is an organized list of primary or other sources relating to a given subject or person. Traditionally, this is the one of the important services a library attempts to provide. Using a Web page, bibliographic details of documents can be presented through user-friendly search facility. This service can further be enhanced by combining it Document Delivery. "Uncover" (URL: http://uncweb.carl.org) is one such bibliographic service with document delivery facility. In a digital environment, this can be highly customized to the extent of making it a SDI service. It should be noted that the present HTML is not suited to present bibliographic data, as they are no tags for the bibliographic descriptive elements. There are two solutions to this problem 1) Adding HTML tags to the output of the library database. This is normally done by CGI (Common gateway Interface) scripts 2) Alternatively, a library can use XML (Extended Markup Language) to define tags for bibliographic data elements, so that entries in the bibliographies can be presented in a kind of catalogue format.

The compilation of a bibliography involves the following steps:

- Definition of the subject,
- Scope of the subject in terms of the coverage to be achieved by the bibliography,
- Items of information to be included for each document,
- Kinds of entries to be prepared,
- Arrangement of entries to be followed,
- Physical form on which the entries are to be prepared,
- Form of bibliography in which it is to be finally to be finally presented to the users.

Some of the libraries compile bibliography on a particular subject only on demand from the users. This section provides details of this service.

Webliographic Service: List of Web documents (list of URLs) available on a specific topic across the Internet can be compiled into a webliography. Here, hotlinks can be given to the documents on the Web, so that the user can click on the links and actually access the Web document. It should be noted that OCLC is doing some research work on cataloguing Internet sources in order to provide content designators to the bibliographic data elements of the Web documents.

Publications: Publications of the library or of the parent institutions can be on the Web page. Many universities having printing press and publication division. If the library has an independent Web page along with that of the institute, a link may be provided to the publication's page. Alternatively, the library can maintain in its Web page information about the university publications.

Supporting Services: The supporting services include

- *Translation Service*

The non-English language potion of the world's scientific and technical literature required by scientists, engineers, and technologists probably amounts to as much as 50 percent. At present, the most expedient way to make this body of foreign language material available to the scientists and engineers of the system is by the provision of a translation facility.

Scope of Technical Translation

The technical translation activity would have the following objectives:

- Establishment of a Translation abstracting service,
- Provision of an adequate responsive and anticipatory translation service,
- Maintenance of a Translation Bank,
- Cover-to-cover translation of selected journals,
- In-service training for building up a cadre of highly skilled specialized translators,
- Establishment of cooperation and coordination with International Information Systems.

- *Reprography Service*

The art and method of document reproduction broadly defined as "Reprographic Methods", have come to stay as a basic necessity in every aspect of modern life, more so in the field of communication and dissemination of information.

The advantage of reprography has found innumerable applications in information centres and their scope is practically unlimited. The advantages are:

- Procuring copies of out-of-print material or unpublished works such as laboratory notes;
- Securing the contents of valuable documents;
- Publication in limited editions;
- Saving storage space;
- Acquisition of copies of portions of books, periodicals, etc. without having to acquire the whole of the original etc.
- *Printing Service:* Apart from providing access to documents by the supply of copies, the reprography unit would also be bringing out the abstracting periodicals, current awareness services, cover-to-cover translations and other information publications. It may also be required to bring out the organizations Technical Journal, Conference Seminar and Symposia papers, Proceedings etc. Therefore, printing facilities are required to be set up. In a computer-oriented automated environment the lithography system can make use of computer printout directly.

Instruction Service: This service can also be called as User Education. This service aims to provide knowledge and skills necessary to find out his own way. The objectives of user education are:

- A general orientation to available facilities and resources
- The teaching of basic research skills and strategies and
- The teaching of the organization of the literature in various disciplines, as well as basic reference tools in each discipline.

Thus the achievement of the objectives will facilitates recognition on the part of users of their own information needs as formulation of those needs. The user education constituted of the four interrelated components. These are:

- User awareness,
- Library orientation,
- Interest profiling,
- Bibliographic instruction.

Once users have been made aware of the library as a primary source of information, then they must be oriented of library facilities. Thus user awareness is the initial stage, preceding library orientation. Interest profiling should precede bibliographic instruction. Library orientation and bibliographic instruction form the basic components of user education. Library orientation precedes bibliographic instruction. Library orientation is supposed to lay groundwork for bibliographic instruction whereby the users can become effective users.

Bulletin Board Service: Very often, researchers and academicians need to discuss their individual findings and study with fellow academicians.

The quickest and easiest way of doing this would be to have a type of Discussion Forum or Bulletin Board Service on the Internet, which would be a kind of bulletin board for members to "pin their notes" for others to see. Generally Bulletin Board Service is used to invite people's interest in special resources and services: to house interests in selected books on a particular topic; to present short, graphic summaries of important national and local news; and to announce programmes and lectures. The bulletin boards should be installed in places where too many readers come.

Special Collection Service: The Special Collections Unit of the university library will have its mission to provide access to unique research collections. These collections of primary and secondary resources are acquired, organized, and preserved by the library and are made available to qualified researchers. All the book materials in the collection are catalogued and accessible through the online catalogue. Access to the original records in the archives, collections of personal papers, business records, political papers, oral histories, and maps is provided by indexes and finding aids in the unit. Any qualified person doing scholarly research is permitted to use material housed in the Special Collections Unit.

CD-ROM Service: CD-ROM is a high capacity storage media, which is widely used in libraries due to its various advantages. Libraries store different reference books as they take lots of space on the shelf and also because of their heavy weights. Different abstracting and indexing journals are now available on CD-ROM, for example, LISA (Library and Information Science Abstract), ISA (Indian Science Abstract), etc. The libraries those have infrastructure to use CD-ROM provide this service to the users. It is a way of providing bibliographic services using CD-ROM. Another reason of libraries using CD-ROM is that some documents are available in CD-ROM. To give access to those books libraries provide CD-ROM service.

Online Book Suggestion: This is a good service usually offered by the university. One can suggest a book for the library to acquire via e-mail. For this the person in charge of this matter should have e-mail facility. The user should provide the following items of the documents to be acquired.

- The name and department of the user who is suggesting.
- The status of the user (*e.g.* undergraduate, graduate, faculty, staff) with e-mail address.
- For each item suggested, one should provide Author (required), Title (required), Publisher, Year of publication, Price, ISBN number.

Access to Internet and E-mail: This is a new and good service and those libraries provide it that has Internet access. It allows the users to access Internet and get the relevant information of their interest. It also allows communicating with others by providing access to E-mail.

Through e-mail renewal facility, the library can open reservation facility. If users are familiar with Internet then they can be allowed to access Internet. Otherwise library staffs help them to get what they want.

Links to Other Library Sites/ OPACs: It is not uncommon among the library Web pages to provide link to other nearby or important libraries. It becomes rather mandatory if the library is a participant of a library network.

Intranet Implementation: All the library services and the information about it need not be provided to the entire Intranet community. Some of the facilities and services can be made a perogative to the faculty and the students of the parent institution. In such a case, Intranet on the campus wide network (LAN), can be used to make such information available to the local users.

LIBRARY ONLINE RESOURCES

Library Online Catalogue

Generally all university library having Online Public Access Catalogue (OPAC). Here information about the OPAC should be mentioned. Search faclity through Subject, Title, Author, Author/Title, Call Number, Subject Keywords, Title Keywords, ISBN/ISSN Number etc. should also be there. Reserve facility by course name, by faculty member will add the advantages of the catalogue. Facility of getting indexing and abstracting service can help in researcher, so it will add the value of the online catalogue.

Online Databases

A database is a collection of related data stored in a central location or in multiple locations. Libraries provide access to their databases through their Web pages. The catalogue of a library informs the users about the books or journals, which contain the information of their requirement. To get the original information, one has to access the electronic journals or books. The online databases provide access to journals and books online. Some of databases are restricted to the users of a particular library and some of them are available for others also.

Some databases are password protected and some are free. It may need to pay the cost to access a database or it may be free also. Generally, the databases are arranged by subject. The databases can include references sources like Proquest Direct Research Library, OCLC, Encyclopedia Britannica Online, SilverPlatter Online, CARL Uncover, ERIC etc. All these reference sources contain thousands of different types of database and mostly these are free. It will be very useful to the ordinary users and also researchers.

PROQUEST DIRECT RESEARCH LIBRARY

Proquest Direct Research Library: ProQuest gives instant access to one of the most extensive collections of published material in the world. One can search through thousands of journals, periodicals, and newspapers in Bell & Howell Information and Learning's vast archives. To bring that information to the desktop, ProQuest provides a unique mix of article searching, information access, and information delivery. ProQuest helps professional researchers and students find what they need quickly, conveniently, and economically.

OCLC: OCLC is a nonprofit, membership, library computer service and research organization dedicated to the public purposes of furthering access to the world's information and reducing information costs.Researchers, students, faculty, scholars, professional librarians, and other information seekers use OCLC services to obtain bibliographic, abstract, and full-text information when and where they need it, in a form they want, at a cost they can afford. OCLC helps libraries provide affordable electronic access to a rich range of information for users— abstracts, full text, indexes, and bibliography—whether the user is in a library, office, or home.

Encyclopedia Britannica Online: Here's opportunity is there for 30 consecutive days of free access to the 32-volume Encyclop)ia Britannica—plus thousands of newly added articles, graphics, and related Internet links. Through membership anytime one access this source.

SilverPlatter Online: SilverPlatter is the leading provider of electronic information to researchers and librarians worldwide. Their solutions allow librarians and knowledge workers in academic, medical, and corporate libraries to offer seamless desktop access to the highest quality research information. It includes

Research Databases

They publish over 250 premium electronic databases covering a broad range of disciplines.

Internet Service: The Internet Subscription Service provides hassle-free access to SilverPlatter databases via the Internet.

Search and Retrieval Software: SPIRS (SilverPlatter Information Retrieval System) is the most widely used cross-database search and retrieval software in the world.

Networking Software: ERL Technology provides efficient access to critical information from any location, across any network or Intranet, on all major platforms.

Full-Text: The SilverLinker Database integrates bibliographic references and the corresponding full text articles, wherever they exist.

MEDLINE

MEDLINE (MEDlars onLINE) is the National Library of Medicine's (NLM) premier bibliographic database covering the fields of medicine, nursing, dentistry, veterinary medicine, the health care system, and the preclinical sciences. The MEDLINE file contains bibliographic citations and author abstracts from approximately 3,900 current biomedical journals published in the United States and 70 foreign countries.

The file contains approximately 9 million records dating back to 1966. Coverage is worldwide, but most records are from English-language sources or have English abstracts. Each MEDLINE record is identified with a unique identifying number called a MEDLINE UID (MUID in PubMed). Citations for MEDLINE are created by the National Library of Medicine, International MEDLARS partners, and cooperating professional organizations. MEDLINE records are incorporated into PubMed weekly, and are also assigned a PubMed unique identifier (PMID).

CARL Uncover

UnCover is a database of current article information taken from well over 18,000 multidisciplinary journals. UnCover contains brief descriptive information for over 8,800,000 articles which have appeared since Fall 1988. UnCover offers the opportunity to order fax copies of the articles from this database. UnCover is easy to use, with keyword access to article titles and summaries.

UnCover Reveal is an automated alerting service that delivers the table of contents of your favourite periodicals directly to the e-mail box. The Reveal service also allows users to create search strategies for their favourite topics.

ERIC

ERIC, the Educational Resources Information Centre, is a federally-funded project of the U.S. Department of Education, Office of Educational Research and Improvement. The ERIC database is an internationally-accessible collection of practically- and research-oriented documents for educational practitioners and researchers at all levels. In addition to all these there can be links to the electronic databases which can be arranged subjectwise. A union list of serials can be created to facilitate the users. A guide can be provided to the users about how to search the databases. Inaddition to this subject-oriented users' guides and the general guide about the sources of information can save the time of the users.

Course Assignment Guides

The Course Assignment Guides page provides online guides to Library sources designed to assist students working on assignments for particular courses. These lists are prepared by Reference librarians in consultation with the faculty teaching the courses. They are displayed for the length of the semester, then removed.

General Interests

This particular topic contains information of general likings. Recent topic, burning topic or topic of general interests can be put here. Some of the general topics are How to Evaluate Information on the Web

Writings Tips

Copyright

The year 2000 Problems

Internet Research areas

In addition to all these any university can put some Internet Tutorials site also.

How to evaluate information on the web:

The World Wide Web has tons of information, but not all web sites are equally valuable or reliable. A researcher must be able to evaluate information on the Web. Web sites must be critically evaluated. Some of these criteria are:

Evaluating Website is a Snap:

S = Source, Somebody created this site? Who?

N = Nature, Why does this site exist? Purpose?

A = Appearance/Accessibility, Does this site function efficiently?

P = Page Content, Is the information accuate and reliable?

Some other thing should be borne in mind, these are:

1. *Content Specialist:* Does the site cover the topic comprehensively? Accurately?

Can anybody understand what is being said? Is it written above or below your level of understanding?

What is unique about this site? Does it offer something others do not?

Are the links well-chosen? sufficient?

Currency: Can that website tell: the date the information was created? the publication date? the date the material was last revised? Are these dates meaningful in terms of the subject matter?

Would you get better information in a book? an encyclopedia?

Would you include this site in your bibliography?

2. Authority/Credibility specialist: Who is responsible for this site?

What are his/her credentials?

Have the authors of the site cited their own sources?

What is the domain name? Does it end in.com,.gov,.edu? Is that a meaningful clue in evaluating the site?

Would you include this site in your bibliography?

3. Bias/purpose specialist: Why was this site created? Is it a personal, commercial, government or organization site?

Is there any bias? Is only one side of the argument presented? Is there a hidden message? Is it trying to persuade you or change your opinion? Is the bias useful to you in some way?

Can you distinguish facts from opinion?

Would you include this site in your bibliography?

4. Usability/design specialist: Is the site easy to navigate (user-friendly)?

Is there a well-labelled contents area?

Do all the design elements (graphics, art, buttons, etc.) enhance the message of the site? Is there consistency in the basic formats of each page?

Are there any errors in spelling or grammar?

Do the pages appear clean, uncluttered?

Do the links on the site work?

Check whether the website included in the bibliograohy?

Use evaluation criteria based on the needs. So, in sumary we can conclude that the evaluation criteria depends on:

Authority of the Source

Who is the author?

Author's credentials, occupation, experience?

Balance, objectivity

Does the author represent an institution or organization?

Will author affiliation bias the information?

Currency

Is the information up to date?

When was the page last updated?

Site Organization

Is the site easy to use?

Are formats and speed acceptable?

Purpose

Who is the intended audience?

Is the purpose to inform or persuade?

Are there advertisements on the page?

Compare with other sources

Are other sources better (periodicals, books, etc.)?

Is there a cost for the service?

Writing Tips

The following web sites can give the full description about this writing tips. These are:

Rules of Writing: This site is a concise guide to some of the most commonly violated rules of writing. It is intended for all writers as an aid in the learning and refining of writing skills.

Common Errors in English: This site addresses deviations from the standard use of English as judged by sophisticated users such as professional writers, editors, teachers, and literate executives and personnel officers.

Elements of Style: Online version of William Strunk Jr.'s 1918 edition of Elements of Style. Includes elementary rules of usage, principles of composition, elements of form, and commonly misused expressions.

Online English Grammar: This site is can give a alphabetical list of grammar problems, practice tests, and a grammar clinic to post individual problems.

Common Style Manuals

APA Style

Bibliography Styles Handbook

This Handbook summarizes and illustrates the bibliographical formatting rules for three different citation styles: the American Psychological Association (APA) style, the new Modern Languages Association (MLA) style, and the old MLA style.

How to Cite Information from the World Wide Web

This site giveHow to cite documents retrieved from the World Wide Web, APA style.

Psychology with Style

APA hypertext guide to writing research reports for students and instructors of experimental and research methods courses. Summarizes the Publication Manual of the American Psychological Association (4th edition).

MLA Style

Bibliography of Writing Styles

This Handbook summarizes and illustrates the bibliographical formatting rules for three different citation styles: the American Psychological Association (APA) style, the new Modern Languages Association (MLA) style, and the old MLA style.

The Columbia Guide to Online Style

Provides examples of MLA-Style citations for FTP, WWW, telnet, and gopher sites as well as synchronous communications, e-mail, and newslists.

MLA Style

Guidelines on MLA documentation style, authorized by the Modern Language Association of America.

OTHER STYLES

Chicago Style-Works Cited, or References

Includes examples of citing books, articles, encyclopedia entries, dissertations, reprints, government documents, papers presented at conferences, performances, videocassette recordings, interviews, and secondary sources in Chicago style.

Documentation Guide-Turabian

Basic introduction to citation style for the 6th edition of Kate Turabian's A Manual for Writers of Term Papers, Theses, and Dissertations (Chicago: University of Chicago Press, 1996).

General Guidelines for Citing Government Publications

Hints for citing government publications; maintained by the Government Documents Department at the University of North Texas Libraries.

Copyright Basics

This site answers the following questions: what is copyright; who owns the copyright; what can be copyrighted; how long can a work be copyrighted.

Use of Copyrighted Materials

This site discusses how copyrighted materials may be used and the guidelines in place governing this use.

Fair Use

This site describes the Fair Use statute that allows the use of a copyrighted work without permission and discusses the four factors to consider when applying fair use.

Distributed Learning

Explores electronic reserves and what can and cannot be copyrighted in the distributed learning environment.

Web Copyright

This site examines the issues of caching, linking, DOIs, and metatag infringement.

Resources

Contains links to useful Web sites dealing with copyright.

The Year 2000 Problem: Internet Resources

Overview

The Problem: Until recently, the date in virtually all software programmes and hardware components has been represented by two digits instead of four, in order to save computer memory. In the year 2000, a computer or software programme that can't tell whether the date "00" is 1900 or 2000 might either crash or misinterpret the data. This potential failure has come to be known as the Year 2000 (or Y2K) problem.

This situation is compounded by the leap-year problem. By definition, every year divisible by four is a leap year, except for years divisible by one hundred, which are not leap years—except for years divisible by four hundred, which are leap years (an exception to the exception). A computer system that uses the same representation for 1900 and 2000 may not recognize that February 2000 has 29 days.

Y2K Compliance: Definitions of what it means to be Y2K-compliant vary depending on what industry, organization, or software vendor you consult. Most companies and government agencies, for obvious reasons, will be extraordinarily optimistic and reassuring in estimating their own level of compliance. First Chicago NBD (which recently merged with Banc One to form Bank One Corporation) has been working on the Y2K problem since 1995. Their Year 2000 site lists the following criteria for compliance:

General Integrity: No value for the current date will cause any interruption in operation.

Date Integrity Date-based functionality must behave consistently for dates prior to, during, and after year 2000. Explicit/ Implicit Century. In all interfaces and data storage, the century in any date must be specified either explicitly or by clear logic techniques such as windowing.

Leap Year The year 2000 must be recognized as a leap year, and systems must process the date February 29, 2000, properly.

Note that even if a company or agency's own equipment and software are compliant, that company or agency (and its clients) could still be adversely affected by the Y2K problem if its suppliers or the infrastructure (transportation system, communication system, etc.) are not compliant.

Consequences of Non-Compliance: Identifying and upgrading all systems and equipment that represent the date with two digits are enormous jobs.

Computer systems with the potential of crashing or malfunctioning are found in virtually every modern convenience that our current society depends on, including elevators, heating and air conditioning systems, ATM machines, credit cards, cash registers, automated medical equipment, cars, telephones, fax machines, copiers, cellular phones, security systems, vault doors, radar systems, traffic lights, power stations, water and sewage systems, telephone exchanges, and television cable systems. The

Y2k problem is aggravated by the close interconnections among information systems in today's society. As mentioned earlier, failure in one system could have serious impact on others.

Since nothing like this has ever happened in the history of the world, nobody really knows exactly what the consequences of the Y2K problem will entail. Predictions range from a few mild and temporary inconveniences

to a total breakdown of civilization and the end of the world as we know it. The truth probably lies somewhere in between. We do know that the something must be done to avoid serious consequences, and the solution will undoubtedly prove to be enormously time-consuming and expensive.

The U.S. National Institute of Standards and Technology (NIST) estimates that about 167,000 lines of code per staff-year can be analyzed, modified, and tested. This means that large corporations can expect to spend between $5 million and $10 million on corrective action. The Gartner Group, the world's leading independent advisor to business professionals making information technology decisions, estimates that Fortune 500 companies will spend between $10 million and $40 million each. Worldwide, the figure is $300-600 billion.

Achieving Compliance

Companies: NIST recommends the following procedure for organizations and companies to achieve Y2K compliance:

1 Select a product to assist in managing the inventory of software and databases involved. Select one or more products to assist in analyzing the software and estimating the extent of the problem. Some of these products will also modify the software and data automatically, but cannot do so for every case. (Some computations are date-related, but cannot be determined from the source code. In such cases, an individual must analyze the source code line by line.)

2 Inventory applications, libraries, databases, extraneous files, documentation, and other items that have importance within specific systems. Identify who is responsible for each item.

3 *Analyze the Applications and Data*: Estimate modifying the source code alone to change those locations that perform date computations and logic operations based on dates. Perform a second estimation that includes modifying databases and all source code that references data fields and all source code affected in the first estimate.

If there is an insignificant difference between the two estimates, the recommended course of action is to modify both the databases and the source code. It may be less expensi ve in the short run to modify only the source code, but more expensive in the long run if maintenance problems crop up over time due to the date processing fixes.

4. Assemble a team of programmers, application experts, database designers, and project management based on the overall system requirements. Once estimates are known, the number of personnel required can be determined, particularly in view of the automated tools selected for use.
5. *Modify the System*: Three major options are: a) modify the source code to manipulate and perform computations on dates with century digits included; b) use a sliding window time frame to determine date context for computations; and c) incorporate packed date fields and use specialized subroutines for performing the computations. All three of these are expensive and may lead to further maintenance problems in the long run.
6. *Test the Modifications*: Allow 40-50 percent of the overall project resources for testing, even more if the database is modified. This includes testing documentation to ensure that directions are correct and correspond to the changes made.

WEB-BASED SERVICES EXPECTED FROM LIBRARIES

Academic scenario, over the years, has undergone a tremendous change assuming new dimensions influenced by the technology driven applications. Management education is no exception to this. Traditional commerce and management education methods are observed to be inadequate. Quality service package delivery is a formidable task for all institutions of business education (Gupta, 2003). The long-term sustainable advantages in the business education require more attention to the issue of service, quality and cost in the national and international markets.

Libraries are mainly entrusted with a host of predetermined tasks like acquiring, organizing, preserving, retrieving and disseminating information to the users. Right from ancient times to the present Internet era, the primary objective of library has always been this. However, the way this purpose has been achieved has drastically changed. Information technology has influenced the very nature of business and management libraries.

They are undergoing significant changes today not only in outlook but also in function, services, methods and techniques for collection development, processing and dissemination of information (Singh & Krishna, 2004). The conventional set up of brick and mortar libraries that store information within a constrained physical space have given way to data centres that integrate data sources around the globe by way of networking. Libraries have not yet explored their full potential to the full (Miao, 2001). With the advancement in technology and its direct application to libraries, business and management libraries are becoming lean and agile libraries that streamline information supply. The pervasive nature of the Internet, coupled with platform independent database connectivity is turning library portals more and more effective.

The main purpose of this study is to study the availability of websites at management institutes and the extent of library information hosted on it. The study also aims to examine the reasons behind why websites have failed get the attention of majority of users and to identify the library services that they wish to carry through the internet.

WEB PORTALS AND LIBRARIES

A Web Portal can be defined as a website for a specific audience that aggregates an array of content and provides an array of services (Pienaar, 2003). They are the sites on the World Wide Web that typically provide personalized capabilities to their visitors.

They are designed to use distributed applications, different numbers and types of middleware, and hardware to provide services from a number of different sources. Content linked in library portal is superior to the open access content available on the Web (Joint, 2005). Authentication software, commonly known as Web Access Management (WAM) are available that allow the library to govern the access to licensed electronic content (Myhill, 2004). Commonly referred to as simply a portal, it is a website that offers access to a broad array of resources and services of libraries such as e-journals, online databases, Web OPAC, new additions and any other static information about library services.

The developments in information and communication technologies (ICT) and their subsequent absorption in library and information science (LIS) have forced information professionals to change the way they are functioning at present. Because of their popularity with the users, an overwhelming attention is being given to the web-based information services in libraries (Krishnamurthy & Chan, 2005).

LIS community has realized that the academic world is increasingly becoming web-based. Those institutes who wish to attract prospective students and the faculty members must show significant improvements in the institute's infrastructure. Libraries with mere books and journals may no longer lend a credible support to the academic fraternity.

Having a tech savvy library and information centre with exhaustive information resources in all formats has become unavoidable. Libraries are procuring more and more electronic sources like electronic journals, electronic books, online databases along with locally digitized theses and dissertations. The efforts of libraries in providing users with an integrated way of checking the availability of a source in all possible formats have necessitated a properly designed web portal (Cox & Yeates, 2003).

Management libraries are also supposed to define and redefine their services and continuously keep their efforts on value addition to the services offered by them. Proliferation of electronic resources has posed several challenges like multiple logins, multiple interfaces and resource discovery. Web portal offers effective solutions to these challenges (Groenewegen & Huggard, 2003).

NEED FOR THE STUDY

Information and communication technologies (ICT) have brought a lot of conveniences to the library users (Gupta, 1996). Information being accessible at their desktop, the productivity of the students and faculty members has increased. As libraries continues to be flooded by more and more electronic resources, future students will become more and more dependent on them (McGeary, 2005). Libraries in business and management institutes use propriety or in-house software packages for library management.

Even though there is a growing trend in libraries to move towards e-resources, some of the libraries still depend upon books and journals. Print sources continue to dominate the library environment in such cases. In this scenario, basic library services like online public access catalogue (OPAC), reservation, accessing current awareness bulletins, pay for the library dues and document delivery needs to be done at the convenience of users.

A literature search on the LISA database and scanning the journal articles, conference and seminar proceedings revealed lack of availability of such literature which studies the web-based services expected from libraries in the management institutes. Hence, this need for the study.

OBJECTIVES OF THE STUDY

The objectives of the study are:

- to study the availability of institutes' websites and the extent of library information hosted on the websites;
- to study the regular use of the Internet by users and their frequency of visit to institutes' websites; and
- to identify the library transactions that users wish to carry out using the Internet.

HYPOTHESES

The hypotheses of the study are:

- Institutes have failed in attracting the library users to their websites as they provide only static information about library and its services.

- The majority of the users in management institutes use the Internet on a regular basis.
- Most of transactions users do at the library should be carried out through the Internet at their convenience.

SCOPE AND LIMITATIONS

The scope of the study is limited to studying the web-based services expected from libraries in All India Council of Technical Education (AICTE) approved management institutes affiliated to the University of Mumbai. The limitation is that this article covers only the day to day library transactions like online catalogue, reservations, etc. Access to online resources is not studied as it is understood that they will have to be web-based only.

METHODOLOGY

This article is based on part of a questionnaire survey conducted for the doctoral studies at Karnatak University, Dharwad. This survey covered the management institutes in Mumbai City, Maharashtra State, India to understand and measure the impact of information technology applications on library resources and services. In all 24 management institutes approved by All India Council of Technical Education (AICTE) and affiliated to the University of Mumbai (Appendix I) were surveyed. Different questionnaires were administered to librarians and users (faculty members and students) to understand the usage pattern of the Internet and the web-based library services that are expected. The questionnaire was designed to be quick and easy to complete. Among others, the questionnaire specifically contained questions on:

- Availability of institute website
- Extent of library information on the website
- Regular use of the Internet by the users
- Frequency of visit to the institute website
- Web OPAC availability
- Library transactions wished to be carried out using the Web.

The questionnaire was piloted with few select management institutes and users and their comments were incorporated into the final version.

For better results, questionnaire method was also followed by paying personal visits to the library so that the researchers can discuss with the librarian and users at the institutes. Out of 24 management institutes, 22 institutes responded to the survey by returning the questionnaire, thus resulting into a response rate of 91.66%. The collected data has been duly analyzed and presented below.

Availability of institutes' Websites

Management institutes, new or old, pay particular attention to design and host their own website to showcase and create awareness about their profile.

In this age, having a suitably designed website has become as important as their physical building itself. It is found that all the 22 institutes who have responded to this survey have their own website.

Extent of Library Information on Institutes' Websites

Table presents data about the extent of information available about library on the institutes' websites. It is observed that almost all management institutes 20(90.91%) have only general information about library on their websites.

Extent of Library Information on Institutes' Websites

Extent of Information on Website/Portal	No. of Management Libraries
General Information	22 (100.00)
Library Rules and Regulations	2 (9.09)
Library Staff Details	0 (0.00)
Current Awareness Service / Selective Dissemination of Information Bulletins	0 (0.00)
OPAC	0 (0.00)
Any Other	0 (0.00)

Regular use of the Internet by users

Faculty members and students were asked about their Internet usage. Table mentions that an overwhelming number of faculty members 52(86.67%) and 175(92.11%) students have indicated the use of the Internet on a regular basis.

Regular Use of Internet

Regular use of Internet	Faculty Members	Students	Total
Yes	52 (86.67)	175 (92.10)	227 (90.80)
No	6 (10.00)	14 (7.37)	20 (8.00)
Not sure	2 (3.33)	1 (0.53)	3 (1.20)
Total	60 (100.00)	190 (100.00)	250 (100.00)

It is also observed that 6(10.00%) faculty members and 14(7.37%) students indicated that they do not use the Internet regularly. A negligible 2(3.33%) faculty members and 1(0.53%) student are not sure about their Internet usage habit.

Frequency of visit to institutes' websites

As an Institute Website/Portal is an access point for all resources and services, it is necessary to know the frequency of users' visit to the website. Table records the data about it.

Frequency of Visit to Institute's Website

Frequency of Visit to Institute's website	Faculty Members	Students	Total
Daily	7 (13.46)	19 (10.86)	26 (11.45)
Weekly	9 (17.30)	35 (20.00)	44(19.39)
Fortnightly	18 (34.62)	60 (34.28)	78 (34.36)
Monthly	18 (34.62)	61 (34.86)	79(34.80)
Not at all	00.00	00.00	00.00
Total	52(100.00)	175 (100.00)	227(100.00)

It is observed that about 7(13.46%) faculty members and 19(10.86%) students visit their institutes' websites daily. Nine (17.30%) faculty members and 35(20.00%) students prefer to visit weekly and about 18(34.62%) faculty members and 60(34.28%) students visit the website fortnightly. About 18 (34.62%) of faculty members and 61 (34.8%6) of students visit the website on a monthly basis. The poor number of visits that the institute website receives is because not all institutes give any dynamic and useful information on the website. These websites contain some static information about history of the institute, faculty profile and some description about resource units like library and computer lab.

Web OPAC Availability: Online public access catalogue (OPAC), when made available on the Web is called Web OPAC. Information may be available in any media; it is the catalogue that ensures maximum exposure and subsequent use of information sources. Web OPAC's have provision of boolean operators. While searching library collection using OPAC, we can use operators like AND, OR, NOT and make our query more specific. Web OPAC's also provide facilities like personalized onscreen display, save options, etc. It is surprising to know that none of the management libraries surveyed has its catalogue available on the Web.

Library Transactions and the Internet

Users were asked for an explanation as to what are the library transactions they wish to carry out at their convenience through the Internet. The observations are given in the table. About 40(66.67%) faculty members and 140(73.68%) students indicated that they wish to search books' database. Thirty five (58.33%) faculty members and 116(61.05%) students indicated they wish to know the availability of a particular document. About 47(78.33%) faculty members and 158(83.16%) students, highest number of users, wish to do a refined search of books, using all possible approaches. Over 33(55%) faculty members and 140(73.68%) students indicated that they wish to reserve a book while 38 (63.33%) faculty members and 123 (64.74%) students indicated that they wish to access current awareness bulletins.

Library Transactions Wished to Be Carried Out Using the Internet

Category	Faculty Members	Students
To search the books database	40 (66.67)	140 (73.68)
To know the availability of a particular document	35 (58.33)	116 (61.05)
To search books using all approaches	47 (78.33)	158 (83.16)
To reserve a book	33 (55.00)	140 (73.68)
To access Current Awareness Bulletins	38 (63.33)	123 (64.74)
To pay library dues	11 (18.33)	54 (28.42)
To pay for fee based library services, if any	11 (18.33)	63 (33.16)
To know more about services rendered by library	41 (68.33)	150 (78.95)

About 41(68.33%) faculty members and 150(78.95%) would access the website to know more about services rendered by library. Relatively low preferences were given to activities like to pay library dues; to pay for fee based library services, if any. A close scrutiny and comparison of responses reveals that faculty members and students agree with each other on what transactions they wish to carry out using the Internet.

FINDINGS

- All the management institutes have their own website. However, the institutes' websites make a cursory mention of the library as a resource unit and provide only general information about library. Very insignificant number 2(9.09%) of libraries have provided library rules and regulations on the website. Important details like library staff details, Current Awareness Bulletins/Selective Dissemination of Information (CAS/SDI) bulletins and OPAC is not made available on any institute website. Observations confirm the first hypotheses that the institutes have failed in attracting the library users to their websites as they provide only static information about library and its services.
- The majority of faculty members 52 (86.67%) and students 175 (92.10%) use the Internet on a regular basis. Observations from table confirm the second hypotheses that the majority of the users in management institutes use the Internet on a regular basis.
- In spite of the majority of users being regular users of the Internet, most of them do not visit institute website regularly. Their visits are scattered and range from daily to monthly.
- It was observed that users wish to carry out library transactions like searching for a document with all possible approaches, reservation, access library bulletins, pay for library dues and to know more about library resources and services. Observations from table confirm the third hypotheses that most of transactions users do at the library should be carried out through the Internet at their convenience.

SUGGESTIONS

The survey and the subsequent analysis of the data and the findings of the study have enabled the researchers to provide some practical suggestions about starting and improving the web-based services expected from management libraries.

The main suggestions for improvements are as follows:

- Websites should be developed into a more informative portal with detailed information like staff details with contact information, library bulletins like new additions, CAS and SDI bulletins, etc.
- Libraries should make the portal interactive by hosting the Web OPAC, announcements about new library resources and services, links to the websites of interest to the business and management studies.
- Library portal should provide a facility for users to access library databases and put a claim for document reservation.
- Every effort should be made on a consistent basis to update the library portal frequently. This will certainly entice users to library portal and then to the library resources and services.

CONCLUSION

Developments in information and communication technologies (ICT) have a profound impact on every sphere of academic activity. Library and information management is not an exception for this. Business and management libraries have so far not adapted the innovations of Internet and networking to the fullest extent. Earlier, card catalogue was called as the *"Mirror to the Library"*. Now in this age of the Internet, it is the library portal that is being called as the *"Mirror of the Library"*. Library portal reflects the strengths and weaknesses of the libraries very effectively. They are also the tool through which libraries are trying to reach out to the tech-savvy user. Libraries should make consistent efforts to provide web-based services to their users. The survey has revealed that management libraries in Mumbai city are lagging behind in exploiting the full potential of the library portal. It is hoped that libraries at the management institutes will attend to this lacunae by developing a fully functional web portal as soon as possible.

APPROPRIATE LIBRARY BEHAVIOUR

Print the Appropriate Library Behaviour Policy

By entering Carnegie Library of Pittsburgh you are making a commitment to act courteously toward all other persons here;

act respectfully with regard to public property; and follow all rules of this facility. Enjoy the library and allow others the same opportunity. PLEASE observe the following:

- Keep all valuables and personal property with you at all times. The library is not responsible for lost or stolen items.
- Supervise children at all times.
- Follow library computer policies.
- Use cell phones in designated locations, such as the lobby or outside.
- Engage in activities associated with the use of a public library.

Customers not reading, studying, using library materials or attending events may be required to leave the building. In order to provide a safe and comfortable environment, the following are NOT PERMITTED in the library:

- Destruction or theft of library materials or property
- Use of alcohol or illegal drugs
- Disorderly, disruptive or boisterous conduct
- Threatening, harassing or intimidating language or behaviour
- Outside food; uncovered beverages
- Smoking and use of tobacco products
- Sleeping
- Weapons
- Bathing or shaving in public restrooms
- Solicitation of any type
- Pets or animals, other than authorized service animals.

Customers are expected to maintain an acceptable standard of personal hygiene. Creating a hazardous or physically offensive condition-including unpleasant body odours that may offend other library customers as well as infested clothing or personal effects-is unacceptable. Shirts and shoes must be worn at all times. To provide an orderly and pleasant public environment, any staff of this building, including security staff, is authorized to determine whether a library customer is not abiding by these or other Carnegie Library of Pittsburgh rules and regulations. The library reserves the right to respond to any and all conduct not expressly set forth herein but which is deemed by library staff to unreasonably interfere with the use of the library by other customers or interfere with the performance of their duties by library employees.

Any customer who violates these rules and regulations may temporarily or permanently be denied the privilege of access to all Carnegie Library of Pittsburgh facilities. A customer whose privileges have been denied may have the decision reviewed by the Library President or Deputy Director.

Appropriate law enforcement authorities will be notified promptly of any unlawful activity. The Children's Department strives to provide welcome environment for all of its users. For the safety and enjoyment of all who use the library, a responsible adult or caregiver must accompany children in the Children's Department. While in the Library, parents and caregivers are responsible for the safety and behaviour for the safety and behaviour of the child.

UNATTENDED CHILDREN

An unattended child is a child of any age who is apparently unaccompanied by an adult. Parents, guardians, teachers and caregivers may not leave children alone or in the care of other children who are unable or unwilling to provide adequate care. Supervising adults must be close at hand.

As in all public places, "stranger danger" is a real concern. Library staff cannot prevent children from interacting with or leaving with persons who are not the appropriate chaperone.

Staff may refer to Allegheny County Children and Family Services for those children who are left unattended in the Library and whose basic needs for food, rest, parental supervision or attention are not being fulfiled. If the Library is closing, at regular time or in an emergency situation, and a parent or guardian of a child cannot be location in the building, the City of Pittsburgh Police Department will be called. The Library is not responsible for any consequences of parents forfeiting their responsibilities.

EMERGENCY CLOSING

It is the policy of the Carnegie Library of Pittsburgh to recognize that on occasion, inclement weather or other conditions may prevent the library from opening or may cause it to close early. The Library will close for weather related emergency conditions whenever the City of Pittsburgh closes during the weekdays. Because of terrain, there are occasions when one or more of the branches libraries must close because of an emergency situation or the direct impact of weather conditions on neighbourhood(s) where the branch is located. Closing the Carnegie Library of Pittsburgh during regular business hours is highly unlikely, but in the event of severe weather conditions call the Library's mini-message board at 412-622-3116 and/or listen to radio station KDKA for announcements regarding closing. Media notification is made each day the facilities are to be closed.

MEETING ROOM

As part of its service to the community, the Carnegie Library of Pittsburgh provides meeting room space in some facilities for use by cultural groups and neighbourhood and civic organizations. Meeting room space is made available by the library "on an equitable basis, regardless of the beliefs or affiliations of individuals or groups requesting their use." (Article 6, Library Bill of Rights). Meetings must be for purposes consistent with the objectives and goals of the Library. These include meeting community information needs, individual enrichment, and the support of the educational, cultural, civic and recreational activities of the community. Use of library space may be subject to the approval of the Deputy Director. Meeting room facilities are available in many of the agencies of the Carnegie Library of Pittsburgh. Groups are invited to request use of meeting facilities by contacting the agency head. Since facilities and furnishings vary from agency to agency, group representatives may arrange a visit to a facility to view the space before booking a meeting. Agency managers can provide meeting room applications and offer further details regarding eligibility, fees, reservations, equipment and other concerns.

6

Making of Libraries and Librarians

INTRODUCTION

The information age is here with many challenges and risks for the information generator and the information disseminator. The challenges before libraries and information centres are manifold. The main challenge will emerge from the expected change in the very nature of future libraries and information centres.

Library Scenario

The diversity of the Indian library scenario is an important aspect for consideration while speaking about a vision for the 3rd millennium, since different library sectors in India are in various stages of development and no single solution or approach will be relevant in the Indian situation.

Indian library and information sector can be divided into five major areas: the National Library sector; Academic library sector; the Special and Scientific library sector, Government library sector; and the Public library sector due to various factors the scientific and special library sector is much more developed and modernized than the other four sectors.

The challenges which Indian libraries and librarians will encounter in the 3rd millennium are manifold but the most important challenge will be the change in the very nature of libraries and librarians *i.e.*, the libraries must change from collection oriented institutions to service oriented organisations and librarians have to change from custodians of books and documents to information managers and disseminators. In this changing scenario IT will play the most important role. While taking advantage of the different components of information technology the libraries of each sector will have to ensure the following:

1. take stock of the actual need of the library and its users;
2. bring about change in the functions of the library to make it service oriented rather than collection oriented;
3. resource sharing and networking of libraries;
4. consider both preservation and access as vital components of library service;
5. marketing of information and library services;
6. development of library professionals to cope with the new environment;
7. utilization of management techniques for dealing with the new objectives.

At the decision-making level India will need a National Policy for libraries and information centres within the changed socioeconomic environment keeping in mind the new requirements of the different sectors within the field of librarianship. It will now be worthwhile to analyze the present Library and Information scenario before preparing a vision for the 3rd millennium in each sector of librarianship in India.

The National Library Sector

The National Library sector in India can be divided into two types of libraries. The first one comprises the National Library, Calcutta and the recipient Public libraries under the Delivery of Books Act *i.e.,* Delhi Public Library (DPL), Connemara Public Library, Chennai and the State Central Library, Mumbai. These are the depositories of the printed cultural heritage of India under the Delivery of Books Act.

The depository libraries are a very important part of the National Library sector because India is a very vast and multilingual country with rich linguistic heritage. The 3rd millennium will have to see the cooperation between all these four libraries with the National Library at the apex. This can only be implemented if all these libraries are modernized at the same scale and have compatible systems to ensure resource sharing and networking. As these are financed by the Department of Culture (DoC) of the Government of India (GoI), the standards for modernization laid down by the DoC will be applicable to all these libraries. The other part comprises the National level subject specific libraries such as National Science Library, National Medical Library, etc. In both these parts the National Library stands out as the largest single library which acts as the depository as well as the repository of all published material of India. The 20th century saw the development of all these national level libraries in isolation. The main challenge in the 3rd millennium in the national library sector would be to bring about close coordination between all these national level subject libraries and the National Library of India.

Academic Library Sector

The Academic library sector comprises school and college and the university libraries. While evolving a vision for university libraries for the millennium, the main focus should be on the role of the existing university library in addressing the national issues. Since most university libraries in India work in isolation some information need to be collected and collated before preparing such a perspective plan.

1. Data regarding the actual functioning of the library, its collection, user approach, library facilities, IT application status, library manpower status, etc.
2. A national agency may be identified for undertaking this work.

A major aspect of the new millennium will be cooperation at all levels starting from the local to the global level. Cooperation is a prerequisite for all types of networking for resource sharing. Therefore, the following must be ensured before preparing the plan for linkages between university libraries.

1. Institutional commitment to accept the given measures.
2. Fund and support resource sharing solution.
3. Attitudinal changes in library staff to undertake additional responsibility in resource sharing environment.

4. Training of library staff to understand and appreciate the changed situation.

If these prerequisites are made available the university libraries could plan for resource sharing programmes, apprenticeship in libraries for learning new skills and IT training programmes, consultancy in retro-conversion and in developing IT infrastructure. All these put together will be the basis of preparing a perspective plan and evolve a vision for university libraries in the country.

School and College Libraries

Excepting very high brow private schools, most schools in India do not have a library per se in the present situation. This area of librarianship is very neglected and needs immediate and continuous attention. As regards college libraries, most colleges have libraries but other than very well known colleges in each state, the ordinary colleges do not run libraries of any consequence. Therefore, India faces the challenges of actually preparing a perspective plan from scratch *i.e.*, from collection development to networking through IT solutions, and will need a complete blueprint to develop these two types of libraries during the coming centuries.

India has set up the Information and Library Network (INFLIBNET) under the University Grants Commission (UGC) as the inter-university agency on library modernization to take the initiative and address the issues required to implement the perspective plan for the Academic sector. Each university and its affiliated college libraries also should be part of the overall scheme of development.

Distance Education

India has developed a well organized distance learning system through the Indira Gandhi National Open University (IGNOU) to achieve a high percentage of educated citizens as a prelude to an ultimate knowledge based society. The diversity and geographical distance of India need a strong, well-planned distance learning system for all citizens of the country. To make distance learning worthwhile the information base of the library sector will be equally responsible to be able to cater to the information need of Indians different societal levels. IGNOU has started some unique services for students with the help of IT but unless the rich collection of reading material available in different libraries in India are accessible to the younger generation the aim of distance learning gets nullified. Therefore, the academic sector has to be so interconnected that it will also cater to distance learners.

Public Library Sector

The Public library system in India has developed over the years for more than a century, initially under the patronage of the aristocracy. However, the new independent India which developed a democratic society, recognised the need for public libraries which will work for the common good and will build a strong foundation for a democratic set up.

According to the Constitution of India, Public library development is a State responsibility. Therefore, all states are vested with the mandate to set up libraries within the state at different levels. But due to the diversity in the level of state-wise development, the Department of Culture, Government of India has set up Raja Rammohun Roy Library Foundation (RRRLF) to act as a nodal agency for development of public libraries in India.

Though library legislation is almost a prerequisite, all the Indian states have not yet passed the library legislation. Therefore, RRRLF is also working towards a model Public Library Bill which may be able to encompass all the facets needed for development of public libraries to suit the new changing environment. Moreover, in the rural sector the local governments have been made responsible and accountable for the development of public library facilities through new constitutional amendments. Also, the Central Government is taking well-considered steps to make dissemination of information the main activity in all sectors which deal with rural and social development. The 3rd millennium will see a drastic change which will ensure cooperation and collaborative efforts between the information generating agencies (whether government or NGOs) and the libraries and the information dissemination sector. The government is seriously considering to bring about conceptual change in the rural library sector by coordinating its activities along with the continuing education schemes of the National Literacy Mission (NLM).

The RRRLF has also been given the responsibility of resource mobilization for modernisation of the state and district central libraries, the development of infrastructure and training of personnel. The Department of Culture through the Central Secretariat Library (CSL) has laid down standards for library development software and networking as well as for bibliographic description to ensure standardization both in infrastructural development as well as retro-conversion of library catalogue in machine readable format.

Government Libraries

Government libraries in India were developed mainly during the British period to cater to the needs of the decision-makers and the bureaucrats. These libraries have always been institutionalised within the departments of the government. All these departmental and ministerial libraries have collected government documents mainly of their respective department/ministry and have restricted their acquisition to the need of the departmental staff alone. The Central Secretariat Library (CSL) has, however, worked within a broader spectrum being the main library in the ministerial and bureaucratic set up. During the last 50 years some of the government libraries have been developed into well-organised collections which can cater to an informed clientele. Since government information started being handled by the National Informatics Centre (NIC), the use of information technology became the key word in the government sector. The Department of Culture, which is the nodal agency for library development in India, has taken viable steps towards resource sharing and networking of the libraries under its supervision. This includes the National Library, Calcutta, Central Secretariat Library and National Archives of India library, libraries under the Archaeological Survey of India and the Anthropological Survey of India, etc.

Special and Scientific Libraries

The special libraries and the libraries under Science and Technology group are in a much better developmental stage compared to the other four sectors. A good number of these libraries use current IT products and systems like computers, email, CD.ROMs and on-line storing and retrieval. Libraries and information centres of research institutions such as those under Council of Scientific and Industrial Research (CSIR), Defence Research and Development Organisation (DRDO), Indian Council of Agricultural Research (ICAR), Indian Council of Medical Research (ICMR), Department of Atomic Energy (DAE), Indian Space and Research Organisation (ISRO), Department of Science & Technology (DST) and Institutes of higher learning like the Indian Institutes of Technology (IITs), Indian Institute of Science (IIS), Indian National Science Academy (INSA), Indian Institutes of Management (IIMs), are modernized and have been given the necessary resources for development.

Library Networks

Other than INFLIBNET, there are a number of networks which have developed in the past decade. These are either city networks or activity subject specific networks. Developing Libraries Network (DELNET), Calcutta Library Network (CALIBNET), Madras Library Network (MALIBNET), Management Library Network, etc., are important and useful ventures for networking and resource sharing. Future of Library Development and the Challenges in the 3rd Millennium The details given above actually picturizes the latest situation encountered by Indian librarianship till the end of the last millennium.

The main challenge in the Indian situation is to visualise the present situation and bring about changes so as to nullify the observations made in the World Information Report, 1997-98, which are as follows:

- The bulk of the population is not information-dependent in day-to-day work and living.
- A large proportion of the population cannot consume information, especially if it is delivered in written form.
- The vast majority of the population does not have the means to access information.
- People in general are not accustomed to pay, cannot pay or are unwilling to pay for information. In fact, information does not even figure in their list of wants.

- The existing pattern of economic activities does not favour a growth in information consumption.
- The countries do not have sufficient capacity to invest in infrastructural development

Therefore, the future of Indian librarianship lies in descending from its ivory tower of catering to only a select few and the literate minority and to create a system which will penetrate into the actual information need of the majority of Indian people. So, there has to be a major paradigm shift to deal with the information needs of the vast majority of the population who at present do not have means to access information. Specially, the public library system in the country needs to be rejuvenated and re-modeled to cope with the present situation and the future trends.

There will be need to change the following on a broad-based basis in all types of libraries:

a. Library environment

b. Fragmented and isolated services

c. Organisational structure

d. Library collection

e. Information based infrastructure

f. Information and knowledge management

The first step towards this direction has been taken to a certain extent as the Report of the Working Group of the Planning Commission on Libraries and Informatics for the 9th Five Year Plan 1997-2002 addresses some of the challenges which confront Indian librarianship The Challenges and Risks for Librarians and Information Specialists in the 3rd Millennium The biggest challenge for Indian librarianship is to bring about attitudinal change among both library staff and users. Libraries and librarians are still the lowest priority in the decision making process and the librarians are the least visible persons. If this main challenge is confronted by librarians and libraries in real earnest it is only then that one can expect all the necessary changes within the system itself.

Libraries will also have to cope with the initiatives developed in the society in the changed IT environment. Libraries can no longer afford to remain institutionalized passive spectators. All the activities will now have to be tailored to give long distance and often home delivered information.

Opportunities and Risks

The new millennium will open up unprecedented opportunities for information professionals. Employers will no longer look for the traditional career requirements but a flexible work force which can take on different roles and responsibilities as and when required. In this context the following changes are to be brought about in India:

1. Change the teaching curricula in the field of library and information science.
2. Utilize the traditional expertise and apply them effectively in the new situations by understanding the actual requirements and applying these skills.
3. Content generation to suit the new fields of study with the help of IT.
4. Play the role of information managers by utilizing the new technology to give access to the rich resources available in libraries.

Tomorrows information professionals will be knowledge navigators instead of information collectors. In the Indian situation the new generation of librarians will have to face these challenges and risks to remain in the information business in all the sectors. Inter-changeability and capabilities to adjust from one sector to the other will be the biggest challenge for Indian librarians.

National Policy for the New Millennium

To fructify all that has been stated above there is a felt need to frame a National Policy of Libraries and Information Systems and Services, suited to the new environment. The DoC had framed a National Policy on Library and Information System (NAPLIS) during the late 1980s which is a very good foundation for preparing a new policy which will encompass the futuristic viewpoints for the library sector as a whole. It has to spell out all relevant issues, challenges, threats and opportunities which are envisioned for the new millennium.

India is on the threshold of major changes which will be possible only if a worthwhile and modern information system is set up at the right time, with the help of right infrastructure and is administered by the right type of qualified people.

DESIGN AND DEVELOPMENT OF AN INTEGRATED UNIVERSITY LIBRARY

A library plays a pivotal role in ensuring the success of higher degree of research. The important activities of university libraries include the Collection Development, Reference Service circulation, Document Delivery, User Education, Access to Electronic Resources etc. University libraries are expected to provide cost effective and reliable access to information using the state-of-the art information technology tools. Information technology has revolutionized the information handling activities in the academic libraries during the past few years.

The Information Society demands that all the relevant technologies; that are involved in information processing, consolidation, repackaging and retrieval be merged so as to evolve an integrated system; capable of providing diversified services. In this direction the automation of individual university libraries is a first step, rather a prerequisite for the development of such an integrated university library and information system. The promising trend in the development of information services with effective networking of these libraries will facilitate the optimum utilization of information resources.

Indian universities constitute one of the largest higher education systems in the world and today in our country there are more than 318 universities / institutions, 16,500 affiliated colleges, 10 million students with 5 lakhs teachers. This vast academic community needs a wide variety of information services in the changing academic environment. The electronic resources, which are available in libraries today is an outcome of the advances in both computer technologies, with powerful computers the information storage and delivery mechanisms, such as CD-ROMs and user-friendly interfaces.

In most of the academic libraries in the western countries, Online Public Access Catalogues (OPAC's) have almost replaced card catalogues, offering enhanced search capabilities for accessing the local collection and often expand coverage to include the holdings of other area or regional libraries. Many libraries also provide a web interface to their library and information system. The library and information system with a web interface often includes direct links to electronic journals, books and Internet resources.

Access to electronic journals and full-text data is another important component of an integrated library system providing access to full-text resources in an electronic library setting. The consortia models help to provide better access to scholarly literature. The access to electronic resources enable the researchers what they want, when they want it, where they want it. Full-text electronic resources offer access to resources unrestricted by either location or library hours.

Higher Education System

Higher Education has undergone a radical transformation in the last few decades both in UK and the US. Education is one of the most crucial services in the economy. The quality of life of society at large and individuals in the society largely depends upon the quality of education. Higher education institutions in the country especially the Universities have been producing the required quality manpower as per demands made in the social system. Since the dawn of independence in India, higher education system has grown enormously. It has expanded in an unprecedented manner not experienced by any other nation in the world in recent times. The advent of Information Technology products like computer and Internet have further made it possible to spread the higher education even in remote corners of the country (Kamal, 2002).

The higher education in India provides information on degree and diploma awarding universities, institutions of national importance, deemed universities, and colleges imparting general, technical and professional education. In the Indian system, higher education includes the education imparted after the 10 + 2 stage – ten years of primary and secondary education followed by two year higher secondary education. The first degree, the Bachelors degree, is obtained after three years in normal case for arts, science, commerce and four years in the case of professional degrees (four and half in case of medicine and five / six in case of law). The Master's programme is usually of two years duration. The postgraduate programmes except engineering involve 2 years of study after first degree, the M. Phil. Programme is of one-and-half year duration and is a preparatory programme for doctoral level studies. Ph. D programme is research study for 2 years, while D. Sc. and D. Litt are awarded by some universities after Ph. D for the original contributions. (Universities handbook, 2002). During the last 53 years we have very rapid expansion of Higher Education. The Table – I shows the growth of higher education in India (Dongaonkar, 2002).

Over the last few years, large number of private and deemed institutions attained university status and India's higher education system has continued its growth and has reached more than 318 universities at present. Deemed-to-be-universities (also referred to as Deemed Universities) are institutions that are conferred status of a university by virtue of their long tradition of teaching or specialization and excellence in a particular area of knowledge. To cite an example Tata Institute of Social Science Mumbai, Deccan College of Post Graduate and Research Institute, Pune etc. The "Deemed to be University" status is granted by the University Grants Commission (UGC) with the approval of the Department of Education, Ministry of Human Resources Development (Educational Consultants Inda, Ltd. 2005).

Status of Automation and Networking of Universities

The main purpose of this research study was to understand the possible areas where the application of Information Technology (IT) is made as a part of overall improvement in the university library services, and also to understand the pace of accepting the emerging IT particularly Internet in the university libraries. The study also focuses on implications arising out of IT applications, the strategies and future plans to formulate an integrated information system. The data is collected from various sources. The questionnaire was designed for this purpose. It is a tedious task to cover the entire set of universities all over India. Hence the study is restricted to approximately 25% of the total set of universities taking into consideration the regional, types and status of universities. However universities for the study are randomly selected and the details are given below.

A pilot study was conducted to test the suitability of the questionnaire. 16 libraries were covered for pilot study and the response received from 10(63%) libraries were analyzed. Further the responses were critically viewed and certain ambiguities were resolved to draw the final questionnaire. 54 university libraries forming 81.8% of a total population covered are received. The statistical package for social science (SPSS) package was used to feed and tabulate the data. The analysis is done using the frequency distribution technique and the details are given in the following paragraphs.

General Information: It is an established fact that most of the Indian universities have not filled in the post of Librarian though the posts are sanctioned. The survey shows that 48(88.9%) of the libraries reveal that the libraries are headed by library science professionals and some of them are at Assistant Librarian level, where as 6 (11.1%) libraries were headed by the Library In-charge from other subjects.

Qualification of the Librarians: The data shows that 26(48.1%) libraries are having qualified librarian with Ph.D. in Library Science. Where as 17(31.5%) libraries have the professionals with Master's degree in library science. There are few librarians with M. Phil qualifications.

Manpower: The data regarding the manpower available in various universities was collected to know the strength of the competent staff to handle new information services. Since there is hardly any recruitment in number of universities the staff strength has reduced considerably. The data reveals that 22(40.7%) libraries have less than 10 qualified professionals. There are few libraries where even less than 5 staff members working to serve the users. The staff strength in 18(33.3%) university libraries ranges between 10-20. 12(22.2%) libraries have more than 20 qualified professionals. With the implementation of Information Technology the computer science professionals are recruited at various universities. The University Grants Commission-UGC, INFLIBNET supports the appointment of Information scientist. It gives the financial assistance to recruit a person with computer science background or Library science with P G Diploma in computer applications. It is observed from the table that only 27(50%) libraries have computer professionals.

Availability of Infrastructure: Many universities procured computer systems with Pentium – III when they received funds under Automation and networking facility of INFLIBNET and many have not been able to upgrade their computer systems due to financial limitations. However few major and small universities have upgraded the computers time to time and have latest computer systems.

Software Used: Generally the Softwares are used for database development and management of library house keeping operations and word processing activities. Many libraries initially started using CDS/ISIS for data creation etc. but later on they purchased library software. Few universities did not respond to this question however many universities were using SOUL software developed by INFLIBNET and equal number of universities were using commercial software *viz.*, LIBSYS, SLIM, LIBRIS, ALICE and few have developed their in-house software.

Database Development: Database development and maintenance are the most important activities to facilitate optimal resource sharing. Databases created by the libraries according to their requirements. However for this study data in respect of only books, serials, theses and reports are collected. The development of the databases in the libraries is not encouraging at many universities however few universities have made exceptional progress and automated all their activities. Different types of databases are created by these universities *viz.* books, serials, thesis, technical reports, standards, patents, manuscripts etc.

Networking of Universities: Libraries initially had network connectivity using data networks such as NICNET, VSNL, ERNET, and many other private ISPs. These universities had limitation of funds and the networking was not that effective. With the implementation of UGC-Infonet universities provided funds under UGC have received the Internet connectivity ranging from 256 Kbps to 2 Mbps depending the different criteria set up by Joint Technical Tariff Committee (JTTC) appointed by University Grants

Commission. This has enabled universities across the country to have seamless access to Internet and electronic resources under UGC-Infonet E-Journals Consortium. The other universities *viz.* Technical and medical universities have also upgraded their infrastructure in terms of network connectivity and provide access to Internet and other resources.

Future Plans of Universities: The importance of activities and services in the libraries is based on its present services and future plans. Few libraries already have some facilities *viz.* automation, Campus LAN, member to the network etc. However many other universities do not have these facilities and feel important in the present day context. The chart below shows the impact of different activities are found out and ranked. The table shows that very few libraries have the facilities existing with 16(29.6%) libraries already have automated activities similarly the importance of these services and their priority is ranked showing the importance of these factors.

Need for an Integrated Information System

Universities in India have fairly good libraries developed over the years. Due to the technological innovation, lot of changes occurred in day today activities of human being. Libraries are not an exception. Drastic cut in funding, devaluation of currency, user expectations, initiatives from government and other organizations are various reasons for libraries to embark on available information technology. These technological innovations lead to library automation, library cooperation, library networking, resource sharing, use of Internet in the libraries, electronic access to scholarly journals, access to other library catalogues through union catalogues etc.

Most of the libraries have traditionally tried to own resources as much as possible, because owning an item provides faster access to patrons than waiting to borrow or purchase on demand. However the increased cost of maintaining a collection of primary sources and the increased demand for information has resulted in a shift of emphasis from that of ownership to access. In the present situation, as the academic libraries in India have been largely affected by financial constraints in which resource acquisition has been restricted. Most university libraries are ill-equipped to satisfy user needs within their resources with exponential rise in information.

Libraries are unable to continue subscription to many publications due to price escalation and high foreign exchange rates. At the same time there is a conscious duplication of costly library holdings in the absence of convenient sharing mechanism. Scholars in remote areas feel mentally isolated. It is also impossible to fund all libraries to make them self-sufficient to meet the resource requirement are some of the challenges facing the academic libraries. The need for establishing an integrated university library and information system is viewed in terms of the following points

1. Establishment of Integrated system facilitates the enhancement of existing library facilities and also increase accessibility to all other citizens to library resources and services.
2. Can provide training to the library staff who does not currently possess the skills in the use of new information technologies by organizing well-designed training programmes with mechanisms for follow-up technical assistance and support.
3. Promote collaboration and cooperation among libraries for sharing of holdings and technical ability to maximize scarce resources. Limited holdings, financial capacity, and human resources are major barriers to improving library services.
4. Develop recommended standards and guidelines for library services.
5. Increase access to electronic information sources..
6. The cooperative activities are centralized, and they obtain remarkable results in training, the improvement of library Inter-lending, and in publishing a CD-ROM of bibliographical records from participant libraries for effective use.
7. Cooperation is a way to accelerate the evolution of libraries, and to create new services, to facilitate changes, and to save expenses. The library networks are developed to

connect libraries, which ensure the development of union catalogue with locations.

8. In the era of the Internet, electronic documents, and the virtual library, maintaining independent libraries is out of order. In addition, the efforts needed to face the challenges of the information society and the changes that society is demanding of universities are destined to become weaknesses more than strengths in those institutions that face them individually. There are many reasons why it is advisable for libraries to approach these challenges collaboratively.
9. The huge technological opportunities to share information.
10. The high cost of the e-journals.
11. Growing demands of library users.
12. The ultimate goal of cooperation is to join users and the documents with information they need.
13. Consortia represent the possibility to test alternatives to the traditional automated library. They represent the potential to offer the best library services to a wider number of users with all the resources they possess.
14. The successful operation of a library consortium clearly depends on good working relationships among members and between members and the consortium.
15. One of the significant challenges facing academic libraries during times of dynamic change is the ability to understand the needs and perspectives of their users.

Components of an Integrated Library and Information System

Libraries are changing. The old concept of library as a store house of knowledge are giving way to concepts based on development of 'Intermediary' roles in hybrid environment in which the resources are either traditional or in electronic formats. Librarians will need to be very clear about the purposes that their libraries serve, and will need to re-design the range of services they offer with those purposes in mind. All academic libraries virtually depend on the IT systems for their basic operations such as acquisitions, cataloguing, circulation, serials control etc.

The development of IT based systems by organizations with which the libraries deal and within the institution it self has meant that much closer attention has to be paid to the integration of the library's system with others. The functions that are required to provide effective delivery of information requirements need to be integrated. The integrated university library and information system can provide one-stop information services using the state of the art information technology tools. The system designed to serve as integrated university library and information system is expected to cover all the aspects required so that the integrated system can support technologies such as Internet, electronic publications etc. to provide integrated services.

In the context of new millennium, a university's position should be advanced as a leader among the colleges and universities in using the information technology and library services in providing an enriched learning environment. There is a desperate need for a university to make information technology and library services a pervasive and transparent part of the lives of students, faculty and staff. (William Patterson, 2001) The information resources are pervasive when they are available to every one. Those resources are transparent when information, applications and services are available without any delay or limitation of hardware/ software etc. Users must experience information resources as seamlessly integrated into their activities. The integrated university library and information system can provide pervasive access to information resources; to have a greater return with the use of computer and communication tools to return meaningful results for the benefit of research and academic community. This model has to be developed with the following few objectives:

Automate All Functions and Maintain Comprehensive Automated Library System: The library system will maintain an automated library system which supports the internal automation of acquisitions with online ordering; serials control with online claiming; cataloguing with authority control and bibliographic utility interface; circulation with off-line backup, patron telephone and e-mail notification, and telephone renewal; and patron access catalogue etc. Campus networking for connecting all the departments with library and Maintain LAN/Campus LAN / and a wide area network.

Remote sites need to be connected to the central site for access not only to the automated library system, but also to other electronic sources such as a CD-ROM server, an Internet server, an image server, and possible other servers. Because of the bandwidth needs will constantly be changing, a highly scalable technology is required. The WAN will incorporate a LAN in each library facility. Each will utilize Category 5 UTP (unshielded twisted pair) cabling. The topology will be Ethernet. Routers will connect the LANs to the central site via the WAN. Maintain contracts with two Internet Service Providers and regularly evaluate performance.

Contracts can be Maintained with Two Internet Service Providers (ISPs) for Staff and Patron Access to the Internet: The reason for contracting with two ISPs is that it will make it possible to connect no more than half the staff and patrons to any single ISP-something, which is necessary because the service inevitably deteriorates due to the ever-increasing number of users which can outstrip the ISP's capacity. The library will monitor performance not only by soliciting patron feedback, but also by having public service staff log on during known peak periods of activity. The ISP must offer the library rates, which are lower than those extended to individuals.

Seek to Conform to All Relevant Standards: Conformity to all relevant standards is a high priority. All cataloguing will conform to the Anglo-American Cataloguing Code, Second Edition (AACR-2). The database of the automated library system will be developed and maintained in full-MARC format, including bibliographic, authority, holdings, and patron records. UNICODE compliance to be sought to facilitate multilingual user interfaces.

Create all Library Records in Machine Readable form Using Standards: Creation of library database is one of the prerequisites for success of automated systems. Using the relevant standards such as MARC 21 and AACR2 formats the library records are to be created and provide access to library resources. The records so created using the set standards will facilitate easy exchange of records from one library to other library at local, regional, national and international level sharing of resources.

Provide online public catalogue access to within & to campus users and provide access to other library catalogues The access to library catalogue and its resources is done through the user friendly online public access catalogue within the campus or to the outside users. This facility ensure that the users gets an access to information such as the holdings of different type of material in the library of their interest and its availability, shows the status of an item, facilitates the reservation etc. The OPAC will also provide access to information relating to due items with due date etc. OPAC user can also see the status of receipt of latest issues of a scholarly journal in the library. The access to other library catalogue is equally important when the item required by the user is not within the campus or library, he/she may try to access the availability of such items in the nearby libraries where from one can borrow for a limited period using the inter library loan etc.

Provide one or More "Electronic Access Centres" in each Library: The electronic access Centres play an important role in the university system as the end users expect the central library to provide such facility even at nominal cost. The library will provide one or more "electronic access centres"—clusters of PCs which provide access to a variety of electronic publications. The resources accessible through such centres will complement the library's print collection, rather than replacing it. While the most widely consulted electronic publications today are indexes and abstracts, an increasing number of reference publications and full-text/image files of journal articles are becoming available. Each electronic access centre may be configured to also support multimedia access.

Maintain a Web site of its Own: Rather than limiting what is available to remote patrons to the patron access catalogue, the library will maintain a Web site of its own and links to its automated library system, products on its CD-ROM server. The Web server will be configured with a "proxy server" firewall so that those accessing the libraries' automated library system or other servers will not have direct access, but will interact with the firewall, and it in turn will interact with the target system.

Negotiate for Online Reference Services: The library cannot afford to purchase all electronic publications, which may be of interest to its patrons, nor would it want to purchase those not used frequently enough to justify the subscription price. The library will continue to provide access to full text electronic journals, which is less expensive compare to the print archives. The availability of UGC-Infonet Facility is boon for universities.

Provide Document Delivery Service including electronic document delivery Providing document supply and full-text access to online databases plays significant role in the shift from "ownership" to "access". The escalating costs of science and technology journals, budgetary constraints, and availability of science and technology literature via non-traditional sources, such as commercial document supply and full-text online databases, are reshaping academic libraries' science and technology collections, as well as the modes of accessing and delivering scientific information. (Bandyopadhyay, 1999)

ARIEL is a high-speed, high-quality, cost-effective document delivery system that runs on the Internet. Journal articles can be sent from one place to another by scanning the article directly from the journal. Text and graphics are digested into the computer, transmitted over the Internet, and printed on a laser printer at the receiving end. Provide access to academic information *viz.* admission procedure, examination system, evaluation, scholarships, etc. The university library and information system works as model for access to information including the administrative matters etc. The end user should be able to find out the details of the admission procedures in the university system with criteria for selection of students, number of seats in each subject, online application, online results, scholarship details, hostel facilities etc. Though these databases are individually maintained at different places in the university, the integrated system is expected to provide access to such information and updated time to time.

Upgrade skills of the staff by training and orientation time to time on implementation of latest IT tools. Staff must be trained to handle their new responsibilities. As systems become more complex, staff training increases in importance. Training can be limited to small groups to provide both hands-on experience and close monitoring by the trainer. This core group will then train others in the library.

Designate a Full-time Systems Manager: The library will designate a full-time system manager who has responsibility for acting as liaison between the staff and the vendor. Appointment of the system manager ideally occurs before the vendor is selected. The person selected as system manager need not be knowledgeable about electronic data processing but should understand the functions of all of the library's departments and have good interpersonal skills. The system manager will have to reconcile the library's needs with the capabilities of the vendor, coordinate standards development, implement new system features, oversee vendor compliance with the contract, etc.

Phase in Implementation over a Period: The components of the plan can be implemented over a period of time. Reasonable amount of time is required not only for financial reasons, but also because library staff cannot be expected to do everything at once. It also is not practical to develop a plan, which looks further into the future because the rate of change is too rapid. The plan has to be updated each year and a detailed schedule of activities may be drafted. It will include updated specifications and cost figures.

Conclusion

Libraries are changing. The old concept of library as a store house of knowledge are giving way to concepts based on development of 'Intermediary' roles in hybrid environment in which the resources are either traditional or in electronic formats. Librarians will need to be very clear about the purposes that their libraries serve, and will need to re-design the range of services they offer with those purposes in mind. Libraries exist to serve their users, but the user population is increasingly heterogeneous. All academic libraries virtually depend on the IT systems for their basic operations such as acquisitions, cataloguing, circulation, serials control etc.

The functions that are required to provide effective delivery of information requirements need to be integrated. The integrated university library and information system can provide one-stop information services using the state of the art information technology tools. The system designed to serve as integrated university library and information system is expected to cover all the aspects required so that the integrated system can support technologies such as Internet, electronic publications etc. to provide integrated services. The vast information sources which the library gives access to are not only the item held by or owned by the library but also given access to remote information sources and handling the resultant requirements to authenticate and authorize users. These are the key challenges for the modern academic librarian. (Wendi, Arant, 2001) In the context of new millennium, a university's position should be advanced as a leader among the colleges and universities in using the information technology and library services in providing an enriched learning environment. There is a desperate need for a university to make information technology and library services a pervasive and transparent part of the lives of students, faculty and staff. (William Patterson, 2001)

Academic libraries in India have long desired one-stop shopping for their customers and in this electronic age their customers are demanding it to search from a single point at any physical location, and retrieve information from the library catalogue, citations from journal indexes and full text information from electronic resources. University libraries must provide reliable, cost efficient access to information whether print or multimedia and whether held locally or remotely. The need to provide information services that remove the barriers of distance and time become even more important. In earlier times libraries have always acquired and organized material so that the information is accessible more easily. Libraries are an integral part of the academic mission of a university. Libraries can enhance a university's reputation by providing access to World class information resources and services and can stimulate research by promoting collections and services widely.

GENERAL LIBRARY RULES AND CODE OF CONDUCT

1. To become a member, an applicant must fill out the registration form and submit it through proper channel to the University Librarian.
2. The Library is primarily for the use of students, faculty, officers, non-teaching employees and technical staff of the University.

3. Teachers of affiliated colleges of NEHU can also become members under "Other Categories" subject to fulfilment of enrolment procedure.
4. On special request, Students, Research Scholars and Faculty of other Universities/Institutions can also avail the Library facilities for a shorter period for academic purposes with written permission of the University Librarian. An endorsement from the concerned Head of Institution/ Department or Supervisor on letterhead will be acceptable.
5. The privilege of borrowing books from the Library is restricted to registered library members only.
6. Research scholars from other universities/institutes can avail the Library facilities with the approval of the University Librarian. Such temporary membership is for consultation, reference, and photocopying services only.
7. A fine of Rs. 2.00 per day will be collected for a general book kept beyond the due date. A fine of Rs.5/-per day will be realised for a Course Book kept beyond the due date.
8. Loose issues and bound volumes of periodicals are not for issue.
9. A book will only be issued to a Library member against production of Library Passbook/Borrower's Card(s) issued by the Library.
10. The Borrower's Cards of Students/Research Scholars will remain valid for the period mentioned on the Borrower's Cards and must be renewed in time.
11. The University Librarian reserves the right to recall any book from any Library member at any time.
12. The loan period may be shortened by the University Librarian if the books are on special demand.
13. A book borrowed from the Library may be renewed for a further period provided no other reader has reserved the same.
14. Library members are responsible for the safe custody of the Library Passbooks/Borrower's Cards issued to them.

15. Tampering with the entries/records on the Library Passbooks/Borrower's Cards can render library membership liable to suspension.
16. Library members must not cause any disturbance or engage in any behaviour which interferes with Library activities. Proscribed behaviour includes (but is not limited to) verbal abuse, threats of violence, sexual harassment, or use of aggressive words to harass any person.
17. Library members must not carry lethal weapons or other potentially dangerous items inside the Library.
18. Stern action will be initiated against Library members found vandalizing or defacing the Library building, furniture or equipment.
19. Loss of Library Passbook/Borrower's Card(s) should be reported to the Circulation Section and Course-Books Section immediately.
20. Duplicate Library Passbook/Borrower's Cards will be issued after one week from the date of application on payment of Rs. 25/-.
21. Members shall be held responsible for any loss which the Library may suffer through the loss or misuse of their Library Passbooks/Borrower's Card(s).
22. In case of loss of a book by the borrower he/she shall either replace the book or bear its cost. If the book is out of print, it may be replaced with a bound photocopy as per library specifications. If the book of a multi volume set is damaged or lost the member concerned shall be liable to replace the whole set or pay four times the cost of the same.
23. Library Passbooks/Borrower's Cards are not transferable.
24. Library members must not remove or attempt to take library materials, equipment or property without proper checkout or other official Library authorization.
25. Library members must not conceal library materials in the Library for the exclusive use of an individual or group.
26. Library members must not mutilate library materials by marking, underlining, removing pages or portions of pages,

removing binding or in any other way damage or deface library materials.

27. Library members must not enter unauthorized areas of the Library, remain in the Library after closing or refuse to leave the Library when requested to leave during emergency situations.
28. Library members shall be required to pay the penalty imposed upon them by the University Librarian for any damage caused by them to the books or any other property belonging to the Library.
29. Any damage or missing pages detected in a book must be reported to the staff before borrowing the same, failing which the borrower shall be held responsible for the damage/mutilation.
30. Members leaving the library should stop at the exit so that the books borrowed or materials taken out of the library by them may be checked by the Library Gate-Keepers/ Security Staff.
31. Upon any infringement of the library rules members shall forfeit the privileges of admission to and membership of the library.
32. The University Librarian reserves the right to suspend the membership of any member found misbehaving, abusing the library staff or disrupting the peaceful Library atmosphere.
33 The Library membership of any member found tearing pages or caught while attempting to steal books will be suspended forthwith from using the library facilities and further disciplinary action will be initiated by the University.
34. Members are instructed not to bring casual visitors/friends or outsiders for using the library facilities without the prior permission of the University Librarian.

GUIDELINES FOR INTERNET SERVICES

1. Internet facility is restricted to the registered members of the NEHU Library only. No outsiders are allowed to use

this facility. Prior permission of the University Librarian must be obtained by visitors or research scholars from other institutions availing consultation facilities.

2. Library members who need to use the Internet services must produce valid proof of identity.
3. Library members must record their names and other details in the Users' Register before and after using the Internet services. Maximum time allowed per session of browsing is 1 hour under normal accessibility conditions.
4. Library members must make use of the Internet facility only to support their academic and research activities.
5. Library members are requested to help in identifying the unauthorized users since the Library's Internet facility is only for registered library members.
6. Library members are requested to bring to the notice of the University Librarian any misuse of the computer systems.
7. Library members must not intentionally read other users' computer screens, unless invited to do so, as such inappropriate behaviour amounts to breach of privacy..
8. Library members must not monopolize the computer resources, *i.e.*, using more than one computer at a time or spending an inordinate amount of time on a computer. When other library members are waiting, Internet service users are expected to observe the posted time limit.
9. Library members must not install any software not specifically approved by the Library. Word-processing and other application software considered indispensable for academic activities are preloaded on all the computers in the Cyber Room.
10. Library members must not attempt to bypass system restrictions or tamper with system files or applications. Unauthorized access to system files and/or attempting to disrupt the integrity of the system(s) tantamount to violation of Library Rules and Regulations.
11. Library members found transmitting, disseminating, printing or downloading sexually explicit images or sound

recordings will be deemed to voluntarily invoke suspension of their library membership.

12. Library members behaving in an abusive or harassing manner in the use of the Internet facilities provided by the Library will not be allowed further use of the Internet services.
13. Library members must not indulge in unauthorized copying of copyright-protected materials. Library members are expected to observe relevant copyright laws and regulations.
14. Library members must not deliberately waste or overload the computer resources.
15. Stern action will be taken against Library members maliciously accessing, altering, deleting, damaging or destroying any computer system, network computer programme or data.

7

Buildings, Furniture, Design and Planning

AN OVERVIEW

Libraries are object-intensive facilities. Their resources, services, and programmes depend on the installation of certain types of furniture and equipment. Without shelving to house hard copy, there would be no place to put books, journals, documents, and other artifacts of the print world. Without microcomputers or terminals, CD-ROM players, printers, microfilm readers/printers, and photocopiers, it would be difficult to provide online services, CD-ROM information, or hard copies of micro media. Staff need service desks, workstations, and work areas to perform their jobs. Patrons perusing hard-copy resources also need places to sit.

Of course, where patrons sit depends on their personalities and how in-depth their browsing will be. Some people prefer to read or study in an attractive area and others couldn't care less. In any event, lounge chairs and sofas and chairs at tables or carrels are important library items. Few people are willing to stand for more than a few minutes while leafing through a periodical, studying a reference book, or researching a specific topic. Chairs have also become essential aspects of a large percentage of the online public access stations being installed today.

When OPACs first appeared on the library market, library planners believed that patrons would stand while performing quick searches. Although many patrons do not mind standing, many more prefer to sit. Besides, today's terminals are constantly being loaded with host databases. Browsing through these takes such a long time that sitdown stations are showing up everywhere. It is not uncommon, for example, to find a large academic library's reference area outfitted with four stand-up and twenty-six sit-down OPAC stations.

The problem is that each additional chair costs money, and construction budgets tend to disregard this fact. Funds are often encumbered for construction only, and monies for "loose" furniture must be garnered elsewhere. The same is true for electronic equipment (*e.g.*, microcomputers and CD-ROM players) and general supplies (*e.g.*, wastepaper baskets, pencils, paper, and desk sets). The construction budget ignores these completely.

Where plans for construction of new facilities are concerned, knowledge of the architectural contract and the resulting contract documents (blueprints and specifications) is essential. It is imperative to know exactly what these do contain. In some instances, all "millwork" or custom built woodworking is to be designed and constructed under the architectural contract. Millwork of ten includes custom built service desks, built-in display cases, and similar aspects of interior design. On the same project, shelving may also be considered part of the architecture This is often the case on very large installations; for medium to small installations it is not. Funding for shelving falls into the loose furniture category, which also includes all library technical furniture (tables, carrels, chairs, atlas stands, etc.) and office workstations and chairs. Surprisingly, carpeting is nearly always part of the architectural contract because it provides the finished floor.

Sometimes the budget contains all the items necessary to build and operate the facility-construction, loose furniture, supplies, and electronic equipment. The library administration and staff are informed that a certain amount of money is available, and it is up to them to divide the sum logically. If the renovation/new addition comes in over budget, there is less money to spend for other items.

Having enough money to spend on the proper furniture, supplies, and equipment is not enough, however. The idea is to be cost effective and maintain a low overhead once the project is complete. The building must be able to operate relatively efficiently. Here, the design of the interior architecture is extremely important. That is one of the reasons why library consultants are kept on projects beyond the programming stages. They critique the interior architecture and, later, the interior design space plan.

For example, a proliferation of dividing walls promises operational inefficiencies and thus more staff. Walls impede traffic flow which, in turn, forces employees to waste considerable time getting from place to place. Additional floors or more than one entrance also demand more staff. Too many libraries have had to add more service desks/control points-and employees-to prevent security problems.

It is logical to assume that the interior architecture affects any building's space layout possibilities. An old school converted into a library may have long corridors and a variety of cinderblock walls that once delineated classrooms. An award-winning public or academic library building may feature a vast central atrium, "flying" staircases, and many attractive but unusual areas. In both cases, interior architecture is rather inflexible and limits layouts. The spaces that are created within the envelope are usually characterized as fixed function; these tend to resist logical rearrangement.

For example, if a school was designed as a classroom facility, only activities that fit into 400 square feet segments will function properly. Few library collections have logical breaks which enable them to fit neatly into spaces that are just that size. An award-winning building's central atrium can be an important aesthetic. Its primary function is to bring a sense of grandeur to the interior. One can look up and see through to the next story or look down and view the floor below. Unfortunately, a central atrium creates a "ring around the rosey" effect. Patrons and staff must walk in circles to get from here to there.

For the budget conscious, it is important to note that atriums are also nearly as expensive to heat, ventilate, or air condition as the full floors they replace. Furthermore, buildings with atriums are very difficult to balance mechanically. Service calls that require fixing such gadgets as malfunctioning vents, fans, circulators, pumps, and blowers become a constant fact of life.

Filling in a central atrium is always a solution, but it is one thing to tear down the interior walls of a 1950s school building and another to deck over the glorious atrium of an award-winning building. In both situations, the expense may cause a public furor, but the protests are bound to rise to untenable heights whenever political forces believe that bureaucrats are about to destroy a precious work of art. Similarly, if the school building was erected at the turn of the century, it immediately becomes a historic structure. Should it be replete with special details and fine appointments, resistance to any architectural changes could be defended by an equally ferocious political battle.

Old school buildings are not the only historic structures. Libraries with historical significance seem to be everywhere. There are any number of seventy, eighty, and ninety year old structures still functioning, and they house a variety of libraries-public, academic, governmental, and private. These buildings evoke great affection, even those that have not been well maintained and, thus, have deteriorated. Communities may have ignored their existence, but once one of these structures enters the spotlight, it is amazing how many people profess kinship. The populace tends to view the structures as examples of a gentler age and something they wish to return to-even if they were never there.

Indeed, some of these structures feature architectural details that are either too expensive to fund today or literally against the law. For example, old buildings tend to have impressive exterior stairs that were built without regard to barrier-free environments and, of course, do not comply with Americans With Disabilities Act (ADA) guidelines. Sometimes a stair leads to a very handsome entrance flanked by difficult to open heavy wooden doors. Not only are these doors phenomenally expensive to replace, they also are incompatible with ADA guidelines.

Once inside one of these old buildings, the interior architecture and related interior design all too often limit the ability to conduct state-of-the-art library services. An imposing but inflexible teak and granite circulation desk may take up far too much room. In order to add terminals and other details of automated circulation services, makeshift work areas have been created behind and to the side of it. In close proximity to the desk are one or two wood paneled reading rooms whose floors were not constructed to bear the 150 pound per square foot live loads that library bookstacks presently require. Since the majority of the collection was not expected to be open to the public, it was placed in a once closed and now open access metal self-supporting stack whose small entrance is located to the back of the facility. Within the stack, the only access to the second and third tiers is via a narrow metal stair.

The inflexibility of this building's design implies that there is only one way to perform library service-and, at the time it was erected, that probably was the case. Its architect wanted to create an important work of art that could support processes that were clearly defined.

Of course, library services have changed dramatically over the ensuing years. Now their facilities are expected to house a wide variety of activities, some which came into existence just recently, perhaps only yesterday. Indeed, radical changes in library missions and goals are occurring daily, but the buildings that are expected to support these activities are still being designed with century old rules in mind. The result? A host of new structures that are quickly becoming outdated. The situation is so common that library consultants often receive urgent telephone calls late into the night from harried librarians administering inflexible, barely relevant, buildings that are less than five years old.

Until ADA went into effect, any number of new buildings were constructed with the older models in mind. Too many buildings were designed with requisite impressive exterior stairs that led to equally impressive but hard to open front doors. Administrators now find that they must scurry to find places to add exterior ramps or elevators as well as inexpensive ways to install automatic doors. Beyond inappropriate exterior access, another difficulty concerns the all too common confusing internal pathways. First time patrons complain that they cannot find the interior elevators or stairs. It is not uncommon to find disabled users being forced to traverse long distances before they reach the ramp that will lead them up or down a two step level. Of course the most universal inadequacies relate to insufficient collection and user spaces; nearly nonexistent electrical and telecommunications wiring; too few places to install equipment-dedicated seating; and inappropriate meeting, conference, or training rooms.

Because new construction or reorganization/renovation can be costly, it is not surprising that, in an era of tight money, academic, public, corporate, or governmental financial officers resist making any changes at all. Although librarians take it for granted that we are living in a global information economy, arguments may be forthcoming that it is not necessary to upgrade the building. In five years the book will disappear. With dial-in capabilities, everyone will have access to the virtual library. Or conversely, adding substantial electronics to a building is an expensive and unnecessary use of space. Spending money on hardware and software will diminish the book budget.

When money is tight, allocating resources does tend to be a zero sum game. Furthermore, whether books or electronic equipment are more attractive tends to be in the eye of the beholder. While too many funding authorities are finding it increasingly difficult to believe that hard-copy collections are still growing, in this age of high speed data, librarians still find ways to relegate microforms and microcomputers to small enclosed rooms in dreary basements or other dismal places.

The irony is that, while electronics are threatening to chase print-on-paper out of some facilities, hard-copy publications are still proliferating. Everyone thinks there will come a day when hard-copy collecting will come to an abrupt stop, but more than likely that event will occur far into the future. An increasing number of books and periodicals are being published in third world and developing countries, especially in the far east. Scientific subjects are multiplying and diversifying. New medical practices and innovative drugs command individual subclassifications. International law is becoming of interest to the ordinary person in the street. That is why few buildings are being erected without some place to install compact shelving.

Depending on the method of construction, the difference between a floor that has the loading capacity of 150 pounds per square foot live load or one with 300 pounds may only be a dollar or two more for each square foot erected. To minimize this cost on the upper floors, only one floor may be designated for compact shelving. In other situations, a quadrant slicing through the building's floors may have its columns and floors reinforced. In many cases, the most inexpensive method is to place compact shelving on the ground floor. This tactic usually requires only a thicker floor slab-provided, of course, the subsoil can support the weight of fully loaded compact stacks. The rails upon which these units slide can either be a part of the floor slab construction or added later. If the latter is the case, then the floor to ceiling height should be sufficiently high to take the addition that the track assemblies require.

At an overwhelming majority of libraries, an installation of compact shelving appears to go hand in hand with increasing reliance on electronic services. No one wants to stop collecting hard copy, but space must be created in the public service areas for online searching and CD-ROM workstations. After all, online services and local area networks promise to overcome the limits of architecture and, at the same time, put a cap on the number of renovations to be made. Within the telecommunications cabling, there will be streams of data that must be able to pierce ceilings, walls, and floors.

Here, a major consideration concerns the amount of electrical and telecommunications power that is brought to the building from the various utilities in the planning stage. It is important not to be too conservative. In the near future more is bound to be required.

A rough rule of thumb is that each piece of electric/electronic equipment requires five amps. For example, five times sixty pieces of initial equipment amounts to 300 amps, where those sixty include microcomputers, terminal printers, copiers, microform machines, and electric pencil sharpeners-and coffee pots, microwave ovens, toasters, and so on. Do not forget the substantial amount of electricity required to run all the mechanical and electrical building equipment-heating, ventilating and air conditioning (HVAC), and lighting.

In a moderately-sized building, the HVAC and lighting needs may add up to more than three times the amount needed by information systems and workstation equipment. Although the former's requirements may stay static, the latter's will not. The number of electric/electronic devices is bound to keep on growing. It is only a small increase in cost to bring more electricity to the building in the initial planning. Larger cables may be all that is required. Once construction has been completed, bringing more power may require a large addition of money. Stringing cables is a labour-intensive process.

Another consideration revolves around the availability of cableways, ducts, and other aspects of wire management within the facility. Future retrofits can be expensive if horizontal and vertical power distribution has not been planned carefully. It is not necessary to run substantially more wiring than initially needed. Rather, it is wise to plan building details that will allow wiring and cabling to be added sometime in the future. Most people will think twice before they drill into a marble wall or through good oak molding. They will go to lengths (no pun intended) to avoid unsightly wires from being draped from one end of the room to another.

Knowledge of local codes is also important. Some codes restrict how wiring is run in the plenum above the suspended ceiling; ducts must be provided for that purpose. To bring the wiring down, channels may have to be cut in the plasterboard around columns or in walls. To run wiring along the floor, attractive and newly installed broadloom may have to be cut and spliced and the cement beneath chiseled to create trenches.

If the library designer chooses broadloom, then the option of using undercarpet cabling (flat wiring) closes. The fire code allows carpet tile but prohibits broadloom from hiding this form of wiring. Undercarpet cabling is an excellent retrofit device.

Obviously, the best suggestion is to prevent major wiring problems in the planning stage. During the planning process, ground rules should be created that minimize inflexibilities and thus future expenses. Architectural solutions should come first and interior design solutions second. An architectural solution may be a cellular floor and cable trays along upper walls, while interior design solutions may consist of furniture containing wire management. It is essential that these ground rules be followed during the design phases and not jettisoned the first time a schematic is displayed or opposition is voiced.

For example, since carpet tile costs about 20 percent more than broadloom, it is often hard to sell it to the powers that be. It is clearly the better choice, however. Not only can it act as a future retrofitting device, it is also easier to maintain. One can simply lift up a dirty tile and exchange it with a clean one-perhaps from attic stock or underneath a desk. Tiles in very active walkways can be replaced on a regular basis, perhaps every few years, without affecting any other areas.

Other suggestions to minimize inflexibilities concern the shape of the building's interior. Simply shaped spaces lend themselves to rearrangement whereas complicated ones do not. Whenever the spaces are simple, the resulting areas can be used in any number of ways. Complicated spaces, on the other hand, tend to define the activities that can and cannot be performed. For example, a large open area can house books, seating areas, service points, or instructional facilities, of ten by simple rearrangement, but an interior "street" that threads through alternately narrow and wide spaces may force the adjacent square footage to be used only as originally intended-as offices, group study rooms, storage areas, etc.

Another example of important guidelines concerns the roof and the suspended ceiling. Under no circumstances should either be dropped over the main stacks to minimize construction costs. This is a tactic used by many architects. In a single story building, initial costs can be somewhat lessened by reducing the total cubic area to be erected. In a multistory building, by dropping the suspended ceiling and letting the ducts run just above it, less interior space has to be finished which, in turn, minimizes costs. At first glance these tactics appear to have a second benefit-the possible reduction of utility costs as well. There is less space to heat, air condition, or light. Unfortunately, by dropping the roof or the suspended ceiling, spaces meant for human habitation in the public service area are created that are only seven and one half feet tall.

Although this is tall enough to accommodate nearly everyone-except perhaps one or two of this nation's basketball players-it can cause the feeling of claustrophobia. Most of us live in homes with finished ceilings about eight feet high, and we are conditioned to like public spaces with ceilings that are even higher. In a place of public accommodation, seven and one half feet is just too low.

Designers agree but argue that few people stay in the stacks for hours on end and reading areas with taller ceilings tend to be only steps away. But what will happen in the future is the primary concern. More than likely, in five, ten, or fifteen years a percentage of bookstacks will no longer be needed. The materials-perhaps bound indexes or periodical backfiles-will be removed and access to the resources will be substituted with online services or CD-ROM networks or some other form of networked micromedia. How can a library recycle public service space that is only seven and one half feet high?

The same question relates to self-supporting stacks. During the open access heyday, from the early 1950s through the late 1980s, purchases of hard copy grew geometrically decade after decade. To squeeze all this material into buildings with insufficient floor space, self-supporting stacks were installed in libraries all over the country.

The height of three tiers amounted to about twenty-three feet. From slab to slab, even the lowest ceilinged building had two floors with about twenty-four feet. Thus, self-supporting stacks can be found in any number of "modern" buildings as well as those that are nearly a century old.

Typically, these structures depend on uprights that pierce each deck and support the stacks above. To demount even one stack, it is essential to start at the top; to do otherwise would cause the whole structure to fall down. Unfortunately, the space on the first floor is what everyone covets the most. The only way to make that space available but leave the upper tiers of the self-supporting stack intact, is to remove shelves and leave the uprights right where they are. The result is an unattractive area studded with posts every three feet. Because the problem is so endemic, there are any number of libraries that contain at least one such area. Witness seating in an academic library with three foot wide student carrels shoved between the uprights. At more than one major public library, workstations have been installed in the decks, and the staff forced to work in them often complain about the conditions vociferously.

The gist of the foregoing discussion is to avoid creating unpleasant spaces in public service areas. They will affect the library's future ability to function effectively. For small libraries or libraries with very long runs of bookstacks, for example, the floor to finished ceiling height should be a minimum of nine feet, while a better guideline is eleven and a half feet. It not only is less claustrophobic, it also enables better air circulation and light distribution-provided lighting runs either perpendicular to the stacks or is set in a nondirectional pattern on the ceiling. Further, the fire code requires eighteen inches from the top of an obstruction to the bottom of the sprinkler head. Although one can install sprinkler heads that are flush to the ceiling, in the less expensive installations they tend to protrude an inch or two below. This diminishes the required clear space above the stack canopies. In several well-publicized incidents, top shelves had to be removed by order of the local fire marshall. At one famous law school, the library had to move one-seventh of the collection elsewhere. Everyone knows that off-site storage is an expense they would rather not incur.

Lighting: Another suggestion to hold down capital and operating costs concerns the lighting system. Today, there are any number of wonderful modular systems on the market. Fluorescents come in a wide variety of shapes and sizes, some of which are high output and last for six years or more. Electronic ballasts, which do not drip or buzz and provide dimming and features heretofore unheard of, are widely available. Metal halide high intensity discharge lamps are perfect for very high ceilings or for indirect lighting. These can even be installed in fixtures that appear to be antiques. Incandescent lighting, on the other hand, is to be avoided except for some exhibit areas. This form of lighting tends to be quite inefficient. In ordinary, approximately 90 percent of the electrical energy results in heat. One person described an incandescent bulb as a heating device that just happens to create light. Not only is the energy wasted, but it puts a load on the air conditioning system during its season of operation and thus escalates costs.

Planning Facilities: Designing lighting systems is in the purview of an electrical engineer, while designing mechanical systems falls under the jurisdiction of the mechanical engineer. An architect must work with civil, electrical, mechanical, and structural engineers, landscape architects, and cost estimators as consultants unless they are members of the same firm. Those companies employing both architects and engineers are known as A/E firms. Other specialists involved might be acoustical, audio/video, or computers/networking consultants. Somewhere between 40 and 45 percent of the architectural fee is paid to these consultants.

Regardless of whether the project is a renovation, renovation/addition, or new construction, the various phases of design are known as schematics, design development, construction documents, and contract administration. During the schematic phase, the architect presents the design concept for the project. Elevations-two-dimensional representational drawings of the exterior and interior-is usually provided as are block layouts of the interior. A model or three-dimensional drawings may be created, but this depends upon whether or not the project is large and/or they are specifically required by the contract. Drawings provided by the architect during schematics are typically used for fund-raising purposes. A rendering-a representational drawing of the exterior or interior-is usually considered a separate item. Since it is a very desirable item, it should be budgeted.

The gist of the foregoing discussion is to avoid creating unpleasant spaces in public service areas. They will affect the library's future ability to function effectively. For small libraries or libraries with very long runs of bookstacks, for example, the floor to finished ceiling height should be a minimum of nine feet, while a better guideline is eleven and a half feet. It not only is less claustrophobic, it also enables better air circulation and light distribution-provided lighting runs either perpendicular to the stacks or is set in a nondirectional pattern on the ceiling. Further, the fire code requires eighteen inches from the top of an obstruction to the bottom of the sprinkler head. Although one can install sprinkler heads that are flush to the ceiling, in the less expensive installations they tend to protrude an inch or two below. This diminishes the required clear space above the stack canopies. In several well-publicized incidents, top shelves had to be removed by order of the local fire marshall. At one famous law school, the library had to move one-seventh of the collection elsewhere. Everyone knows that off-site storage is an expense they would rather not incur.

Lighting: Another suggestion to hold down capital and operating costs concerns the lighting system. Today, there are any number of wonderful modular systems on the market. Fluorescents come in a wide variety of shapes and sizes, some of which are high output and last for six years or more. Electronic ballasts, which do not drip or buzz and provide dimming and features heretofore unheard of, are widely available. Metal halide high intensity discharge lamps are perfect for very high ceilings or for indirect lighting. These can even be installed in fixtures that appear to be antiques.

Incandescent lighting, on the other hand, is to be avoided except for some exhibit areas. This form of lighting tends to be quite inefficient. In ordinary, approximately 90 percent of the electrical energy results in heat. One person described an incandescent bulb as a heating device that just happens to create light. Not only is the energy wasted, but it puts a load on the air conditioning system during its season of operation and thus escalates costs.

Planning Facilities: Designing lighting systems is in the purview of an electrical engineer, while designing mechanical systems falls under the jurisdiction of the mechanical engineer. An architect must work with civil, electrical, mechanical, and structural engineers, landscape architects, and cost estimators as consultants unless they are members of the same firm. Those companies employing both architects and engineers are known as A/E firms. Other specialists involved might be acoustical, audio/video, or computers/networking consultants. Somewhere between 40 and 45 percent of the architectural fee is paid to these consultants.

Regardless of whether the project is a renovation, renovation/addition, or new construction, the various phases of design are known as schematics, design development, construction documents, and contract administration. During the schematic phase, the architect presents the design concept for the project. Elevations-two-dimensional representational drawings of the exterior and interior-is usually provided as are block layouts of the interior. A model or three-dimensional drawings may be created, but this depends upon whether or not the project is large and/or they are specifically required by the contract. Drawings provided by the architect during schematics are typically used for fund-raising purposes. A rendering-a representational drawing of the exterior or interior-is usually considered a separate item. Since it is a very desirable item, it should be budgeted.

Although many people believe to the contrary, cost overruns may be no fault of the architect. Rather, they may relate to the client's requirements or unexpected difficulties encountered during construction. The client may believe that the architect's cost estimates are not in line with local conditions, or may demand an addition-such as a mezzanine-to the design. When the bids arrive, to the client's chagrin, costs per square foot are much higher than thought. Other unexpected expenses may arise. For example, although test borings were made, an underground stream may be found flowing right in the middle of the site once excavation begins. Extra funds must be quickly found to divert it. An underground stream may also cause the design to be substantially changed even though some construction has already occurred. Redrawing at such a late date will expand the scope of work and cause extras to be paid to the architects, engineers, and contractors.

Extras are to be expected on most projects. For smaller buildings, it is hoped, those that occur are limited in scope. For larger buildings, there is always something-perhaps minor-that requires additional money. For example, people in systems management make a decision to purchase a new library information system and it needs to be wired according to the equipment manufacturer's specifications. Unfortunately, those specifications vary from those detailed on the electrical engineer's drawings. Or the new building inspector refuses to let the compact shelving operate unless additional security devices are installed. Professionals in the built environment design field nearly always attempt to build in contingency monies into their cost estimates. A contingency of 10 percent is considered reasonable, but people interested in the bottom line try to reduce this. Some, unfortunately, try to eliminate it entirely. Extra expenses also occur when the process is slowed by the arcane methods of bureaucracies and ferocious battles-sometimes over personality problems-that often occur in the political arena. For a new building, from genesis in the mind of the librarian to actual opening day may be as short as two and one half years. The average is five years. Some projects have taken fifteen years to be completed. A minor renovation may take a year to a year and a half, while a major renovation will take as long as constructing a new building.

During the predesign phase, long before the architect is hired, a library consultant may be retained to perform site selection or write the building programme. Here, a request for proposal (RFP) must be sent, consultants interviewed, and the work performed. From beginning to end, the time span for this aspect of the process is at least six months. For a new addition or brand new building, test borings must be performed, or the land surveyed, and so on and so forth, which requires other professionals to send in proposals, be interviewed, and then selected and retained. Before an architect is hired, it usually takes at least three months to advertise, interview a sufficient number of firms, choose one, and then sign the contract. Unless fast tracked, the entire design process takes a minimum of nine months to a year. Often it takes longer because approvals must make the rounds. Unless also fast tracked, construction can take a minimum of another year. If problems occur or the building is sufficiently large, it obviously will take longer.

Once the architect begins to work, an interior design firm may need to be retained. Many architectural firms can provide interior design, and some clients prefer to use them because it allows the librarian to deal with only one set of people. Other clients believe that it is better to use interior design firms because they are more knowledgeable about furniture, colours, and textures. After all, they concentrate their efforts in the field.

Just as architects have professional societies, such as the American Institute of Architects (AIA), so do interior designers. Their most well-known society is called the American Society of Interior Designers (ASID) and, similar to the AIA, it confers certification. ASID members know how to interface their work with that of the architect so that lighting falls over tables and carrels in the reading areas and hanging cabinets mount on walls that can bear their weight. One of the first tasks any interiors person should perform is to test the building programme within the building envelope-even if this task has already been performed by the architect or library consultant. All major pieces of furniture should be placed in the plan to make certain they fit. The test acts as a reality check. Does the programme call for more shelving than the building can hold? Is there enough room for the workstations specified for the work areas? Once this is done, the designer can go about the business of space planning all the areas, designating which furniture requires lighting and telephones, electric receptacles, and data utility jacks; visiting the showrooms with the client; choosing the furniture-colours and textures-and ultimately creating a set of interior design contract documents.

For a library, the documents are typically divided into three-library technical furniture, shelving, and office furniture. Sometimes the first two are combined, but this limits the number of bidders. There are times, however, that furniture is not procured by bid but rather by state contract. Occasionally, one vendor will be given the entire job on a cost plus profit basis.

The following table provides sample furniture prices that were actually received for a court library. Unit prices are obtained so that different items can be added or deleted at will.

CONCLUSION

The foregoing price list does not indicate the time spent detailing furniture specifications: height, sizes, surface finishes, upholstery, edging, wire management, drawers, cabinets, and so on. Because the bidding process tends to be inexact and similar products of one manufacturer differ from another, the installation phase is fraught with a variety of potential problems. The winning bid's double face shelving may be outfitted with end panels that protrude an inch too far into each aisle. One range is lost for every six installed.

This is the type of field condition that causes librarians stress. Where similar problems eventually solved by the architects or their engineers generally pass unnoticed, they occur with great fanfare here. Librarians understand the nuances of furniture, and it is just this understanding that can cause the greatest difficulties.

The designer has several ways to deal with the situation. An obvious one is to reject the end panels and withhold payment. Another, perhaps more judicious, tactic is to see if the interior designer can redesign the area so that the dimensions are not critical.

Here the point to be made concerns letting professionals do their jobs. A wise librarian should keep tabs on the entire process-from architectural design through furniture installation once the ribbon is cut and the doors open wide. it is up to the library staff to operate efficiently and maintain a job well done.

8

Library Finance: New Needs, New Models

Turning and turning in the widening gyre The falcon cannot hear the falconer; Things fall apart; the centre cannot hold; Mere anarchy is loosed upon the world. (Yeats, 1952, p. 489)

This article will describe a new budget model for libraries. What follows, however, is not just a budget, for a budget is the expression of some organizational reality. This discussion will be an effort to propose both an organization and the budget that flows from, and sustains, it.

At the outset of writing this article, this author was wary of some large pitfalls surrounding the idea of creating a new budget model for libraries. When the present model began to emerge about a century ago, the process for the distribution of knowledge by means of print-on-paper was established and understood. Today, however, the future of the distribution of knowledge is unresolved. The trend is toward electronic distribution but certain aspects remain to be settled including setting standards and solving copyright/licensing problems. Meanwhile, print-on-paper appears to be unconcerned about its new competition as large numbers of books and journals continue to be produced. The dimensions of the new information environment, therefore, are not yet clear, and proposing a model for the library of the future still requires a great deal of guesswork. In other words, it is too soon to present a full-blown new model of library organization and finances.

Another pitfall concerned the stakes of the game. Talk of new models and new paradigms is easy. It pervades our generation.

It is intellectually stimulating and exciting to contemplate radically new ways of approaching our work, especially if there is little or no likelihood that these contemplations will actually have an impact on our libraries. But suggesting a new financial model that might be taken seriously seemed a different matter. This author could just as easily-perhaps more easily-construct a flawed model that would, if taken seriously, unleash a calamity upon the library world.

To be sure, libraries are not prone to respond quickly to any stimulus, so it was decided that risk of calamity was modest. In addition, it was decided that tweaking a budget here and there does not constitute a paradigm change, and most of our efforts at new models fall within the parameters of tweaking. It is not certain where the dividing line falls between merely adjusting one paradigm and actually introducing another. Allocating more money to collections and less to staff does not constitute a paradigm shift. Selling the library to a commercial agency and buying back library services does constitute such a shift. Whether the model suggested later constitutes a paradigm shift is doubtful. But it is certain that it requires fundamental shifts in our priorities and in our approach to librarianship, and such shifts are needed today.

As an academic librarian, my efforts here may inadvertently be more directly applicable to academic libraries than to others. Some of the themes that appear are common to all libraries, and it is hoped that the model will be of interest beyond the academic library community.

Why a New Model?: Pitfalls not withstanding, the invitation to propose a new budget model was accepted because this author is convinced that such a model is a necessity if libraries are to thrive in these last years of the twentieth century. This necessity arises from an inter-related set of circumstances familiar to every librarian. The most prominent of these is the paradoxically terrible and wonderful assault of computer technology on the information world (for the best recent and comprehensive assessment of technology and libraries, see Cummings, et al., 1992). Everywhere within libraries appears the signs of this ongoing assault. From public access catalogues to paperless cataloguing, little remains unaffected. The signs of computer technology also saturate our professional world, pervading the programmes of our professional associations as well as the library literature. It is simply clear that libraries and librarians are undergoing a transformation. The degree to which they will eventually be changed is often debated, but the fact that they are in the process of transformation is beyond question.

While the positive implications of this technological transformation for the dissemination of knowledge are truly monumental, to libraries under financial duress the costs appear monumental as well. In the first generation of library automation, libraries were generally successful at finding incremental support for the cost of technology, including new staffing capabilities. Now, however, as the technological remaking of libraries proceeds, it is increasingly difficult to secure new support. Thus the costs of new technology compete with existing budgetary obligations, requiring the reallocation of funds. Such reallocation within most library budgets is a difficult process that often damages the morale of staff and reduces services to users.

The challenge of incorporating computer technology into libraries and reallocating dollars to pay for it is exacerbated by a second circumstance-the inflexibility of existing library organizational structures. Our present organizational structures evolved from, and are adapted to, the requirements of a print-on-paper environment. Not surprisingly, as librarianship became increasingly professionalized and marked by specialization, the internal boundaries that characterized our organizations became more rigid. Often we quarreled among ourselves, artificially creating great gulfs between the interests of public and technical services. It became difficult and remains difficult to overcome or cross these internal boundaries. Such inflexibility is not conducive to rapid adaptation of new technologies but instead fosters the continuation of old methods and procedures. In a time of rapid change, the organizational model, therefore, fails in its role of facilitating efforts to keep pace.

The problem with such inflexibility is magnified because electronic information does not conform to the parameters and requirements of the print-on-paper environment. Indeed, it ignores and crosses these long-standing boundaries as if they were not there. Networked databases, for instance, do not remain in the confines of any department nor do they respect our specializations. Thus, being organized to manage paper, it is often found to be difficult or impossible to respond to technological opportunities gracefully. This is a serious deficiency during an age in which technological change is constant, pervasive, and rapid and in which we are confronted with laying the foundations for the library of the twenty-first century. It is the worst of times for us to allow old models to hamper creativity and responsiveness.

The technological assault combines with this organizational rigidity to produce one additional circumstance that necessitates changing the model, namely, an unproductive anxiety. On the one hand, change seems inevitable. Technology is a large train, and libraries are stalled on its track. Something has to give. On the other hand, change within libraries seems impossible. Given the trauma associated with major changes in libraries, few librarians, including library administrators, have the necessary courage to risk it. To do so is to invite a barrage of criticism and defensiveness. At its worst, the result can be a frustrating environment where the timid are threatened and the bold are held back, where conflict is incubated, and self interest is encouraged. In such an environment, individual librarians may feel that their professional destiny is out of control and that, ironically, none of the structures either within libraries or professional associations can do anything to help. This anxiety-induced inertia is particularly lamentable because, as a group, librarians are easily capable of meeting the technology challenge that distinguishes this time of transition.

Describing a Transitional Library Model

What is called for in these circumstances is a budget model designed to accommodate libraries in transition. Since it is too soon to propose a model for the library of the future, perhaps we can establish a model that is particularly helpful during a period of change, a transitional library model (TLM). Such a model does not have to possess a timeless quality but rather must be designed for flexibility, adaptability, and responsiveness to opportunity. The goals of such a TLM include the following:

* Providing sufficient continuity for ongoing print-on-paper functions while recognizing and accommodating the requirements of the emerging electronic information environment,
* achieving organizational stability without developing structural rigidity,
* underwriting a new working environment for the library staff and providing for perpetual upgrading of their skills,
* restoring the reality of and sense of control and self-value to librarians.

To achieve these goals, the TLM would have to reorganize priorities and place emphasis on some new features.

Emphasis on Education and Training

The existing model for continuing education and training (E&T) for those who work in libraries has largely featured learning opportunities outside the library connected with professional and membership organizations. While this model has provided many benefits to the profession, it has focused almost solely upon professional librarians and provided little uniformity or continuity within individual libraries. In addition, it is at best difficult for individual libraries to develop a consistent E&T programme using this model.

But consistent and ongoing E&T programmes are just what we need for libraries in transition. The need for keeping technologically current is perhaps the most visible challenge but is by no means the only one. Almost every other new priority outlined later carries with it an E&T requirement. And these E&T challenges can no longer be focused primarily on professional librarians. Support staff-level personnel are valuable and vital partners in creating the library of the future. Their skill levels, too, must keep pace with the demands of a changing environment if they are to serve the library well in this time of change.

In the transitional library model, each library should provide the opportunity for every staff member to receive education and training annually. While some common themes will characterize such E&T programmes, they should, to the degree possible, be customized to the individual library in order to match and support its vision, goals, and objectives. Except for very small libraries, efforts to establish regular ongoing E&T programmes should take the form of in-house training using a faculty of both internal experts and external consultants. In a short time, such an E&T programme will reap manifold benefits since a work force aware of trends and which is up to date on skills will be more capable and confident in meeting the challenges of the changing environment.

A new emphasis on education and training will not be inexpensive. Some businesses spend hundreds-some thousands-of dollars per staff member each year and require as much as two weeks of E&T annually for each individual on training. They would not long continue to do so, of course, unless the benefits were sufficiently valuable to justify it. Fiscal realities may prohibit libraries from jumping rapidly to a programme of such magnitude, but it is urgent that they begin. Counting the value of staff time spent in the E&T programme, the transitional library model proposes that at least 2 percent of the existing salary budget be allocated to E&T.

Emphasis on New Service Opportunities

Since the existing library model is focused on carrying out well-known and well-defined operations, it is usually no one's specific responsibility to pay attention to new service opportunities or to inquire after new means, methods, and technologies. The result is that libraries as often discover new possibilities accidentally as they do through consistent ongoing efforts. In addition, most libraries also find it difficult, both financially and structurally, to take advantage of new discoveries quickly even when their value is compelling. In this regard, libraries typically submit the best ideas for innovation to federal, state, or private funding agencies and wait for them to complete the, relatively slow funding cycles. Only then, usually a year later, do those hot new ideas receive attention.

The transitional library model proposes that a spirit of investigation and experimentation be fostered in libraries and embodied in their budgets. In industry, this would be equivalent to establishing a research and development capacity. Because technology is sufficiently mature, however, for libraries it is more likely a matter of finding and applying existing technologies than developing new ones. The manner in which this spirit of investigation and experimentation is embodied in libraries may vary, but this author favours dispersing it throughout the library leadership. Library leadership is anyone who leads any size unit in the library. It is important that all individuals with leadership responsibilities be made accountable for investigation and experimentation within their spheres of responsibility.

In addition, it is essential that some operating budget resources be allocated to support their efforts. While support for experimentation will continue to be a major focus for fund-raising, it is too important not to receive some small level of funding from the operating budget. For this important period of change when so many new opportunities rapidly appear and just as rapidly disappear, the TLM proposes that a minimum of 3 percent of the operating budget (inclusive of salaries and capital expenditures) be devoted to this library version of research and development.

Emphasis on User Responsiveness

Perhaps because both librarians and library users were bound to certain limitations surrounding the use of books and journals, our interaction and attention to the interests and needs of user have been attenuated. As the new knowledge environment emerges, however, there is an Opportunity to design a system that accords much better with the desires of users. This opportunity exists primarily because information in electronic form offers many more options for distribution and access than do books and journals. With this new medium of recorded knowledge, we are no longer tied to the limits of location and physical handling that were foundations of the paper-centred library.

The argument to put user concerns at the centre of creating the new knowledge environment may seem obvious. After all, attention to customers has been one of the key themes of the U.S. business world for the past decade. Whether the source was the quintessentially American management advice of Tom Peters or the system developed by W. Edwards Deming that revolutionized post World War II Japan, we were admonished on every side to recognize the importance of attentiveness and responsiveness to customers. It has not been so obvious within libraries, however, until recently, and libraries have a long way to go before user concerns take precedence over our own opinions as experts.

In order for libraries to emphasize user concerns and satisfaction, the transitional library model proposes that user analysis must become a regular part of what librarians do. Elsewhere this author has suggested that user analysis should become an integral part of the work of reorganized reference departments, but wherever an individual library may choose to lodge the responsibility, it must be given emphasis during this time of change. Fortunately, its cost will be modest compared to its benefits. User analysis requires a small level of ongoing support for communication with users through studies, surveys, and focus groups. Staffing costs can, for the most part, be built into existing staff time, though some consulting may be required initially. The total costs for user analysis may be as little as two tenths of 1 percent of the operating budget.

A library in transition must find a way to escape the rigidity and inflexibility of the divisions within the existing library model. The transitional library model proposes that the best way to accomplish this is to redesign the library on the basis of a team approach. There are several reasons that the team approach is a preferable alternative for libraries. First, it may offer the most gentle means of initiating organizational change. If desirable, an initial configuration of teams may be fashioned from existing units, departments, and branches. For libraries that are more advanced in organizational skills, less traditional and more forward looking teams may be established. In any case, evolving a library into a team-based operation can be tailored to the style and organizational prowess of each individual library, thus reducing staff resistance and organizational shock. More than any other organizational principle, teams provide for stability without rigidity.

The second reason for preferring a team-based arrangement is that it diminishes the Perception of boundaries and divisions and allows the library to adjust rapidly to new challenges. It achieves this, in part, by reducing the management layers of the organization. This often eliminates entire layers of managers who, for career reasons, must be concerned about protecting their "territories." As a result, it becomes easier to establish new teams that reach across old boundaries. Such teams are often called cross-functional teams. The easy ability to form and reform teams is a distinct organizational advantage in a time when a host of new problems and opportunities arise that do not fit existing organizational units.

A third reason that a team-based approach is preferable is that teamwork can result in higher productivity, better quality, and greater staff morale. It does so through better use of the talents of individuals and through better synergism within the workplace through the reduction of individual isolation (Rees, 1991, p. 37). It also provides these benefits because it allows and encourages everyone to take responsibility for the success of the operation. The team approach empowers individual staff members at all levels to contribute to the destiny of the organization as never before, and in so doing, reduces the number of those who come and go like zombies without interest, joy, or enthusiasm for the work. Thus the team environment offers a means to put an end to the anxiety-induced inertia referred to earlier by restoring to librarians and support staff alike the ability to get directly involved in working out their professional destinies.

The costs of developing a team environment are limited to the price of education and training and the effort necessary to make the mental conversion from the vertically authoritarian workplace to the team environment. Most of us have known nothing other than the authoritarian environment, so the conversion will require patience, education, and practice. Since the recommendation for funding in the above section on E&T includes the cost of team training, the good news is that moving to a team environment places no additional drain on the budget.

If empowering the library staff through redesigning libraries into team-based organizations is to be most effective, then the teams must also be empowered to manage their own budgets. While there are examples of libraries that practice some degree of decentralized budget management, the predominant model is for library administrators to retain fiscal control. Our present environment, therefore, is vertically authoritarian both in terms of decision making and in determining the use of financial resources. The result is an organizational climate that militates against budgetary change even for good reasons. When administrators propose financial change, there is little or no ownership of, or support for, the proposal by the staff. Indeed, staff are typically uninformed about the budget and threatened by suggestions for change. At the same time, when staff wish to recommend a change, there is little likelihood that they will understand the budget process well enough to make their case persuasively or have it taken seriously. The result is that most library budgets remain virtually unchanged from year to year. Thus, it is not sufficient to empower staff through redesigning libraries into team-based organizations; they must also be given financial empowerment in the team context.

The transitional library model, therefore, proposes that the library's budget be allocated to the teams as appropriate, and that the teams be granted both the responsibility and authority to determine and manage the expenditure of the funds. This will allow teams to make and implement choices to cope with a changing environment. If new challenges lie beyond the financial resources of individual teams, as they often will, teams may choose to pool resources in ways never imaginable within the present model. They may also choose to create new teams, staffing and funding them with human and financial resources reallocated from existing teams.

Paying both for maintaining key traditional library functions and for beginning new functions is a large order. Indeed, just paying for ongoing traditional functions is often impossible in the current financial climate. So achieving a financial environment that can to some degree do both seems more like magic than modelling. Yet it can be done if the new model provides the necessary means and incentives to help library staffs stretch library financial resources further than ever before. What is required is a financial commitment to developing new ways of doing library business, and these new ways must create much greater efficiency and cost effectiveness.

Incorporating the new emphases described earlier also requires new ways of doing business. Signs of these new ways are already appearing in some libraries. One such sign is the movement of some libraries toward management techniques usually identified as Total Quality Management (TQM). TQM is a complex system that comes in several varieties but at its core is an effort to base decision making on hard data, to improve quality continuously, to place user concerns at the centre of things, and to create a team-working environment. Adopting a comprehensive organizational strategy such as TQM offers the advantage and convenience of providing a single system that can incorporate the emphases of this transitional library model. But whether libraries adopt an existing system or develop one of their own, tight budgets in this time of new opportunities make it imperative that libraries change old management habits.

The Transitional Library Budget

It would be helpful if libraries would discontinue comparing budgetary ratios and statistics for awhile. Like the rigidity of our present organizational model, comparative statistics help hold our budgets in an unfortunate stasis. Since quality is typically measured and claimed on the basis of such comparisons, it is difficult, if not dangerous, to abandon competitive, comparative statistics as the justification for setting and defending library budgets. In practice, statistical standings are far more important in our current budgeting practices than is user satisfaction. If we could discontinue keeping the comparative statistics and allow libraries to construct budgets for better reasons, perhaps we would see positive and exciting results.

For this reason, the transitional library budget should have as few standard ratios as possible. It should, rather, encourage libraries to make whatever creative changes are necessary to serve their users with the best services, resources, and technologies possible. Its goal is to facilitate the search for the twenty-first century library. The heart of the transitional library budget, therefore, is its lack of prescriptions for "appropriate" levels of expenditures for any of the major budget areas. In the foregoing description of the emphases of the transitional library model (TLM), it has been suggested: for education and training, 2 percent of the salary budget should be allocated; for new service opportunities, 3 percent of the operating budget; for user responsiveness, 2 tenths of 1 percent of the operating budget; for teamwork, only the education and training costs; for fiscal empowerment, no additional costs; and for more effective management systems, only the E&T costs.

These are the only prescribed costs contained in the TLM. This model proposes, therefore, to transform libraries into flexible organizations adept at coping with change for an annual investment of less than 5 percent of their operating budgets.

There will, of course, be other costs each year, but the transitional library model resists projecting them. Indeed, if the model serves its purpose, the internal budget ratios among salaries, collections, access, computer technologies, and other expenditures, will change proportions each year. If teams, for instance, are empowered to choose between filling vacancies or converting salary dollars to technology support, they will make the necessary choices, and changes will begin to take place.

It is the premise of this transitional library model that the changes which will take place among the major budget ratios in the next few years will favour a reduction in expenditures for personnel and a corresponding increase in expenditures for technology, a reduction in expenditures for print-on-paper, and a corresponding increase in expenditures for the electronic distribution of information. The TLM, however, does not attempt to prescribe movement in these directions. Rather, it seeks to establish an environment in which libraries will have the organizational and budgetary flexibility to rearrange their internal finances as necessary to best serve their users during this generation of change.

CONCLUSION

It is my hope that the organizational emphases of the transitional library model and the approach to budgeting that supports it will provide libraries with the, means to prosper under the technology assault and to develop the ability to adapt to change easily. It is also hoped that consistent programmes of continuing education and training combined with empowerment through the team-based approach will improve the sense of confidence and control among library staffs.

Perhaps libraries that adopt the TLM will cease to look like mirror images of one another and, instead, take new shapes that uniquely serve the differing needs of their users. And perhaps in a environment where differences are permissible and experimentation is common, some library will find the perfect budgetary ratios for the library of the future. Until that time, a transitional library model will serve us well.

9

The Design of Library and Archive Buildings

Before embarking on a study of the construction and equipping of an archive repository, it may be useful to begin with some definitions.

Whilst the definition of *archives* is almost everywhere the same-the whole of the documents produced by an organisation, administrative unit, firm, establishment, even a family or a person, in the course of the exercise of their activities and preserved for reference purposes-the same cannot be said of the definition of an *archive repository*.

In some countries, the public archives are organised under a separate and independent administrative body, with its own buildings, staff and regulations; whilst in others they are, to greater or lesser degree, linked to libraries, museums or documentation services. Sometimes they are closely tied to the bodies producing them (as, for example, in England where some major administrative units retain responsibility for their own archives); more frequently they come under a separate and independent administration once they cease to be in current use (as in France, where the Directorate of the Archives of France is responsible for the whole of the public archives with rare exceptions).

There is yet another difference. Some countries recognise two types of archive repository: repositories solely concerned with preservation ("historical archives"), intended solely for documents of historical interest, marked for permanent preservation; and intermediate repositories, where documents destined to be destroyed after shorter or longer periods are housed. This distinction does not exist everywhere, and indeed it is not always necessary when the archive repository is closely linked to the body where the archives originated.

All these differences in ideas, organisation and operation are reflected in the differences between archival buildings. For example, a budding intended to house exclusively the archives of a province will be different in many particulars from one in which are gathered together, under one direction, the archives, library, museum and headquarters of the learned society of this province. Equally the small budding housing the archives of some local administration will have little in common with the enormous one housing the central archives of a modern state with all its ancillary services.

The differences in ideas about the role of an archive service also have a bearing on the construction and equipment of repositories. Some countries still adhere, more or less, to the traditional idea of a complete separation between "historical archives" and "current administration". Here, the archive service only has contact with historians and, to a much lesser degree, with the general public by way of historical exhibitions. Elsewhere, there will be a close connection between the archive service and archive-producing departments, which may result in the transformation of the repository into an administrative documentation service. The budding will then reflect this function by the different evolution of some of the working areas.

Whatever the type and size of an archive service, every repository must meet certain basic requirements, which are the same in each case and are well known. They are: a) preservation of documents in complete security, hence the need for safe storage areas, protected against fire, humidity, excessive sunlight, insects, rodents, burglars, etc.; b) production of documents to those requiring them, hence the need for space for listing, packing, labelling, catalogues and inventories, search room; c) repair of damaged documents; d) documentary reproduction-microfilm, photocopying, etc.

To these basic needs will be added, in varying degrees, secondary needs: public exhibition of documents; production of documents for educational purposes; moulding and repair of seals-administrative documentation, etc.

Thus, it will be seen how pointless it would be to speak of budding and equipping an archive repository without first of all defining the kind of archive service it is intended to house; in other words, without drawing up a brief for the architect.

Drawing up the Brief: Whether the question is one of planning a new budding or of adapting an existing budding, nothing useful will be achieved without a precise brief. It would be wrong for the archivist to expect the architect to draw up the brief: it is essentially the user's job to do this. If the brief has been well thought out and set down, any errors in its execution will be the architect's fault and it will be a simple matter to compare his results with the brief which was before him when he drew up his plans. If on the other hand, the brief has been negligently or imprecisely drawn up, the architect cannot be blamed for imperfections for which he was not responsible.

The care with which this essential document should be drawn up is, therefore, self-evident. The role of this *Manual* is to help archivists in this task but not to take their place. Only the head of an archive service knows exactly its needs (quantitative importance of the documents in his care, importance of future transfers, number of searchers, etc.). Many factors enter into the reckoning. First among these is the amount of money at the architect's disposal for carrying out the job. There would be little point in drawing up an ideal brief including all the most up to date technical improvements-unfortunately always the most expensive-if the budget for construction is a modest one. The first mark of good sense is to know how to equate the brief with the real possibilities of carrying it out.

The archivist's theoretical brief must be set against the practical facts of the problem. It is then for the architect to say what is and what is not practicable. Once he has advised on and agreed to the draft brief which has been submitted to him, the document is finalised and will serve as a basis for drawing up plans.

Archivist and Architect: Everyone to his own job. The archivist knows the needs of his service; the architect knows the solutions which will satisfy these needs. It would be as presumptuous for the archivist to claim the right to draw up the plans for his future repository as it would be for the architect to ignore the suggestions and the possible criticisms of his client. Ideally then, close collaboration between architect and archivist is indispensable for the satisfactory completion of an archival building.

However, in practice things do not always fall out this way: sometimes the archivist has requirements which cannot be met, or, has only vague ideas about the equipment he wants; the architect, for his part, sometimes has set ideas which he seeks to impose despite the archivist's advice, or, through excess of work on hand neglects to keep in touch with the archivist and draws up his plans without consultation.

What should one do in such circumstances? It is clear that the archivist cannot force his way into the architect's office and make a scene. On the other hand, the architect cannot seek the archivist's advice every five minutes. In cases of sharp conflict, recourse can be had to arbitration by higher authority, usually whoever is financing the work. But such extreme measures, happily, are rare. More usually, in these matters, good manners win the day, the more so since both sides recognise that, in their different spheres, each is equally competent.

Whether it is a question of building or adapting, once the preliminary consultations have taken place the matter is now one for the architect. He knows the latest technical developments and is responsible for the actual works. He is in a position to suggest new solutions and draw the archivist's attention to difficulties created by the brief. Only for the most weighty reasons, connected with the proper functioning of an archive service, should the archivist oppose the adoption of the architect's plans, although necessary modifications and improvements may be incorporated as they are being drawn up by mutual agreement.

Later on another problem, often a source of conflict may arise, that of supervision of the work. It is essential for the archivist to follow the course of the work closely, since at this point he may detect minor errors or omissions which can easily be remedied as they occur, but which become irremediable once the job is finished. Unfortunately, the architect rarely voluntarily invites the archivist to inspect progress and the archivist must take the initiative in visiting the site, where his welcome may not always be as friendly as he could wish. However, he must persevere: an archivist who has not closely and minutely followed the course of the work will have forfeited the right of blaming the architect if anything goes wrong once the job is finished. A weekly visit should be part of the archivist's professional duty.

The Basic Choice: A New Building or the Adaptation of an Existing One?: For at least fifty years there has been much argument as to whether it is better to build new premises or adapt existing ones to house archives.

A brief summary of the arguments in favour of each point of view follows:

(a) *In Favour of Old Buildings* : The cost is usually lower since there are normally no major works to be carried out on the main structure of the building. The advances of the last thirty years in the internal renovation of buildings have made it possible to re-use old repositories by modernising their interiors. There may also be an opportunity of saving ancient monuments (palaces, monasteries, churches...) threatened with destruction by converting them into archive repositories. Lastly, an argument of a sentimental kind, not without its value when it is a question of preserving part of a cultural heritage, the atmosphere of an historic building may seem to be more attuned to the nature of archives than a new, impersonal, characterless budding.

Against these arguments, it can always be argued that, with rare exceptions, buildings erected for purposes other than the preservation of archives only lend themselves to the necessary functional adaptation with difficulty and imperfectly. Also, if, as is almost always the case, the whole interior must be re-adapted and fitted out as a preliminary, then the cost is often the same as, if not more than, that of a new budding.

(b) *In Favour of New Buildings:* The obvious advantage is the greater convenience and flexibility: all the requirements of the brief to the architect can be transferred more easily on to the plans for a new budding than to those for an adaptation of an old budding. The architect will also have the advantage of the choice of budding materials, etc.

As to the sentimental argument, as with all such arguments, it can neither be countered nor supported by logical argument. It is true that, in certain cases, the use of an historic budding in which to preserve archives may be desirable from a psychological viewpoint: the prestige of the budding may reflect on the archive service (as is the case in Paris, for example, where the beauty of the Palais Soubise and the Palais Rohan has contributed notably in bringing the National Archives to the attention of the general public). On the other hand, such a solution is frequently chosen solely on the more or less valid grounds of economy, with all the drawbacks this brings to the functioning and future of the service.

Indeed, the types of building which lend themselves to adaptation as an archive repository without costly and complicated internal work are rare. Usually, adaptations of this kind are inconvenient and defective (the experience of the National Archives in Paris, referred to above,. is instructive in this respect).

A compromise solution, which may prove satisfactory when sufficient space is available, is to put the working areas of an archive service in an old building and to build a new repository block immediately next to it; or to convert the nave of an old church into strongrooms and to build a new aisle for the working areas.

To return to the psychological arguments, it should be added that the use of an old building conveys little prestige on an archive service unless it is a stately building of character and artistic merit; this is not the case of old factories or warehouses, or even prisons, which are too often offered for conversion into archive repositories.

The argument against conversion does not apply to records centres for which, contrary to archive repositories properly called, accommodation which is both economical and improvised will suffice.

Finally, one decisive argument in favour of a new budding should not be ignored: in the world of today, it is in the archivist's best interests not to turn his back on his own times. The general public-and administrators-are only too ready to regard archives as a collection of dirty, useless old papers, A new and attractive building can greatly contribute to overcoming such prejudices. To insist too strongly on the occupation of an old budding, however "historical" it may be, will only give firmer roots to the notion that archivists are incurably wedded to the past. The consequences of such an opinion on the future of an archive service are only too obvious.

Second Choice: Single Building or Multiple Buildings?: In most countries, it is usual to concentrate all the archives under the control of one archivist in one single building. The advantages of this are self-evident: economies in construction, equipment, staff and time, since everything is in the same place and thus fetching and carrying is reduced to a minimum. However, there are drawbacks especially if the archives are important: there is a risk that the building will become excessively big, occupying a large surface area, which is expensive if it is in the centre of a town. On the other hand, in case of disaster all the documents will be destroyed together.

For this reason, some countries have adopted the system of splitting their archives between several buildings, separated to a greater or lesser extent one from the other. Thus the risks arising from a disaster can be spread and too great a concentration in one place, with its effect on site costs, avoided. But one immediately runs up against other drawbacks: increases in staff needed for running these dispersed repositories; complications in internal relations; a growing risk of disorder and bad administration. We have also to consider the question whether "historical archives" and "intermediate records" should be kept in the same budding or in distinct buildings (archive repositories versus records centres), a question which has already been alluded to above.

While no strict rule can be defined for all cases, it is nevertheless clear that it would be absurd to disperse archives throughout several buildings when their total bulk is of a moderate size. If, taken together, the whole of archives and intermediate records do not exceed 15,000 to 20,000 linear metres (for the meaning of this expression, see below, p. 139), one single budding should suffice to hold them. But once the figure of about 25,000 lin.m. is reached, it is almost invariably better to separate archives and intermediate records. Of course this discussion about the advisability of a single building or multiple buildings for archives is entirely independent of the fact that, in important archive buildings, the strongrooms often have to be divided into two or more blocks; and also that working areas often are in a distinct block from the strongrooms.

Third Choice: Choice of Site: Whether it is a caw of a new budding or an old one, of a single repository budding or multiple buildings, the proper choice of site is vital. Many mistakes can subsequently be corrected, but not the choice of a bad site: the whole future of an archive service can be jeopardised by such a mistake.

Unfortunately, the "ideal site" is very difficult to define. Many factors enter into the reckoning, some of them contradicting others.

a) Sites to be Avoided: at least it is possible to define without any hesitation sites which should be completely avoided:

i. sites with intrinsic dangers: land liable to flooding; unstable sites (*e.g.* hillsides, where the sub-soil is of clay and liable to land slides); damp, swampy sites; land liable to the effects of heavy seas; land subject to termite infestation, etc.

ii. sites with dangerous surroundings: sites near factories with a high degree of air pollution, or near installations of high fire or explosive risk (gasometers, petrol tanks, explosive depots, etc.), or near possible strategic targets (airfields, major railway centres, etc.). Such sites should be rigorously avoided. Under no circumstances should an archivist agree to build a repository on a site with these major disadvantages.

There are other types of site which should as far as possible be avoided, but to which the same categorical prohibitions do not apply: sites near sources of noise (busy main road, railway station, factory); sites which would preclude future extensions to the repository either by reason of their limited area or because of town planning restrictions on the height of buildings. Sites which are difficult of access by reason of steep slopes or bad roads should, as far as possible, be avoided. In any case, before deciding on any site whatsoever, the soil should be carefully surveyed and trial bored so as to avoid unpleasant surprises when digging of the foundations begins.

b) *Definition of the Best Sites.* If a repository principally intended for the preservation of historical archives is being built, it will be desirable to site it in the university quarter of the town; even more so if libraries and museums are intended to be near the archives, as is the practice in many countries.

If, on the other hand, the repository is for departmental records, it will be preferable to keep them near the centre of administration, which is usually in the heart of a town. Such a site is a prime essential if an administrative documentation centre is linked with the archives, to which officials of the administration must have easy access at all times.

This all seems very straightforward. Unfortunately, it most often happens that:

i. the same budding must house both historical archives and departmental records;
ii. available sites in town centres are scarce, very expensive and thus unsuitable for large archive repositories;
iii. university quarters are situated well away from city centres. To these considerations should be added the facts that students studying in different university departments, colleges and secondary schools must be able to attend the sessions of the educational service in the repository with ease; and that the frequent handling arising from the transfer and destruction of departmental records in administrative archive repositories is hardly practicable in the centre of a town.

The "ideal solution" then seems to vary, depending on whether one is talking of a small archive service or one on a vastly bigger scale.

In the first case, (repositories of between 15,000 and 20,000 linear metres run) every endeavour should be made to find a site in or reasonably near a town centre, easily accessible but sufficiently isolated to allow vans carrying records to turn round on the site with ease. A surface area of 800-1000 sq.m. should suffice for this size of building and it should not be hard to find a suitably located site of this kind.

If, however, it is a question of a large and important archive service, it would be wisest-as has been said before-to distinguish clearly between:

i. current or very recent departmental records, which will either remain in the offices where they were produced or very near them;

ii. "intermediate" records, for which a records centre would be built or fitted out with fumigating facilities, large accession and weeding rooms etc. on the outskirts of a town;

iii. archive repositories properly so called, or "historical archives", situated near other buildings devoted to cultural activities (such as libraries, museums, universities, etc.) in a well equipped budding, where the headquarters of the service are located.

It is extremely important that if buildings to which the public has access are situated away from the centre of a town, they should be easily accessible by public transport; this should be a prime consideration in the choice of a site.

c) *Is it Satisfactory to House Archives Outside Towns?* Since, in the opinion of many people, archives are dead things, it is often suggested that they can be kept in the country, *e.g.* in an old castle or ecclesiastical building, far from a town, where land is cheap. The supporters of this point of view add that maximum isolation is thus ensured and as a result archives are protected, especially from wartime dangers.

Such a solution should be rigorously avoided. It can only end in the suffocation of the archive service, which will be cut off from contact with administration, historians and the learned public. Except in the case of services of the greatest importance (such as the Spanish national historical archive, whose presence in Simancas alone justifies the existence of a hostel in this small country town lost amid the plain of Old Castille), archives, in general, tend to be under-utilised if they are situated outside a town. Of course, the establishment of "security" repositories in isolated places far from centres of population, intended solely for the storage of archives in wartime, may be envisaged. But such repositories should not be confused with permanent peacetime repositories; no one would think of exiling the Mona Lisa to the heart of the Auvergne mountains, or the pictures of the Metropolitan Museum in New York to the Rocky Mountains, or Tutenkamen's treasures to the middle of the Nubian Desert, under pretext of safety or security. Archives, like paintings, sculptures, and books are a part of a living culture; their place is amongst the people who create and use them.

It is always possible to leave the heart of an archive service in a town centre (offices, search rooms, catalogue rooms, etc., *i.e.* those parts of direct service to the public) and to spread the storage accommodation which is the most cumbersome part, out around the edges of the town. Such a solution is both practicable and acceptable, subject always to two essential conditions:

a. that very regular and rapid communication between the various buildings is ensured by a service vehicle;
b. that in the strongroom block, along with accession, weeding and listing rooms and various workshops, there is an officer exclusively concerned with producing documents required in the other budding. It must be stressed that the division of the service between two widely separated buildings raises delicate operational problems and requires numerous staff.

Finally, specialised types of repositories, *e.g.* for security microfilm, are normally found on isolated sites, offering the maximum guarantee of safety against all foreseeable means of destruction.

d) Orientation of the Buildings. Once the site has been chosen, thought must be given to the possible orientation of the budding (in so far as a choice is possible, which will depend on the shape and layout of the site).

Because of the dangers to documents from excessive sunlight, it is better to avoid giving strongrooms a direct southern aspect in the northern hemisphere or a direct northern aspect in the southern hemisphere.

In areas exposed to winds from the sea carrying humidity and salt or to hot, dry winds, it is undesirable to site strongrooms facing these winds, although their damaging effects can be lessened by good insulation against heat and humidity.

For search rooms and other working areas in general, that orientation which gives the best natural light should be sought.

Fourth Choice: A Building above or Below Ground?: The principal attribute of an archive repository is security, and it is tempting to ensure this to the maximum by burying the strongrooms below ground. Moreover, this solution has another attraction in that it is an economical use of a site, especially in a large town. Finally, protection against the hazards of war is inseparable from the idea of underground shelters.

These various reasons explain why, over the past three decades, many countries have either excavated new or adapted existing underground repositories. This solution merits close attention but it is only fair to point out that, despite the advantages mentioned above, there are two very considerable disadvantages:

i. the safe storage of archives underground requires constant ventilation and air-conditioning, which may very well be extremely expensive unless the land is dry, because of the danger of the infiltration of water and stagnation of air in subterranean areas;
ii. the possible dangers of electricity failures, negligible in the case of a repository above ground, are likely to be serious in an underground repository.

Again, unless an existing underground tunnel is used, the cost of an underground budding with the extensive excavations, lining and digging which will be involved, may well be greater than that for budding on the surface of the ground: economies in site costs will be largely counterbalanced by this expenditure.

If the primary consideration is to provide protection against the hazards of conventional warfare, normal underground construction will suffice (beginning some 3-4 in. under the surface, provided there is a satisfactory protective ceiling); if, however, protection against nuclear weapons is what is wanted, then, all available evidence suggests that extremely deep tunnels will be needed, well away from military targets; consequently, the problem here is very different.

In practice, such underground installations as tunnels excavated as wartime air-raid shelters, disused railway tunnels, mine galleries driven into the rock may be considered for use as archive stores, provided that all the necessary air-conditioning, emergency pumps, ventilation etc., have been installed and are in continuous operation: but such installations are usually at some distance from a town and their use will have the undesirable result of separating the archives from their users, which, as has previously been mentioned, should be avoided.

The location of archive repositories below ground, With a few exceptions, therefore, tends to be unsatisfactory, both from an operational and a financial point of view. It is quite usual, however, even desirable for an archive budding to have two or three underground floors, fitted out as strongrooms or areas of special security. Mobile shelving can be used here. In practically all modern office buildings (*e.g.* banks, headquarters of large undertakings, etc.) archives are stored underground in a perfectly satisfactory manner; but it should always be remembered that if this solution is to work properly, an air-conditioning plant is essential and this is expensive.

As to wartime security repositories, which have been alluded to earlier, they fall rather outside the scope of this present Manual. Every important archive service must have access to one of these in case of serious danger, situated well away from centres of population and military targets and built-or better excavated-to withstand modern weapons of war. The construction of such repositories should be gone into in conjunction with civil defence services in the general context of the protection of cultural property in time of war. Here the matter can be but briefly referred to.

Fifth Choice: A Horizontal or Vertical Building?: It has, for a long time, been traditional to build archive repositories at ground level. This is obviously the simplest solution with the fewest constructional problems. However, the size to which major archive services have grown means that the surface area required is now considerable. To take an example: a repository with some 20,000 lin.m. of shelving of the conventional (non-mobile) type will have some 2,600 m. of gangways and passageways, a figure corresponding to the shelving of a medium sized repository. For a larger service, the figure would soon be 6,000 m., 8,000 m. or more.

It can thus be seen that it will be in the interests of the service to reduce gangways, and thereby surface area as much as possible, thus ensuring economical use of the site, which will almost always be costly in a town.

For these reasons, one is led more and more to think of constructing high-rise buildings. With modem architectural techniques these no longer present any difficulties and repositories of 20, 30 and even 40 stories are perfectly feasible. Here, most internal traffic is vertical, by way of lifts, thus eliminating the difficulties of manoeuvring trolleys along apparently interminable gangways.

It should, however, be noted that high-rise buildings need special foundations, which are generally very expensive. There is, therefore, no point in building "archive towers" as a matter of course; the solution to the problem depends on the price of land, the importance of the archives to be stored, the nature of the soil (a tower costs less to erect on rocky ground than on soft), etc.

The working of the lifts should also be borne in mind: in the event of power cuts or break downs, the situation can become catastrophic; and it is by no means easy, outside a big town, to maintain and repair such sensitive lifts as those in a 20 or 30 storey building.

Sixth Choice—Traditional Framework or Framework Incorporating the Load Bearing Uprights of the Shelving as an Integral Part of the Structure?: The use of metal shelving makes it possible for the architect to link together closely the framework of the shelving and that of the building itself, by using a building structure suitable for load bearing shelving.

The respective advantages of a traditional framework and one able to accommodate load bearing shelving are studied below. But it should be emphasised that it is at the stage of drawing up the plans that this choice will have to be made. In effect, a framework to accommodate load bearing shelving requires, at all levels, a network of metal uprights in more or less close-set ranks, which governs the whole internal arrangement of the building. If working areas and strongrooms are placed one above the other, these uprights can cause serious functional problems, and, as a result, the whole plan of the building requires careful thought.

General Plan of the Building: A multipurpose archive budding is composed of three (or four) basic elements:

1. Storage areas (strongrooms);
2. working areas for the staff;
3. areas open to the public (the offices of the director and his assistants fall between these two categories);
4. (possibly) official residence.

There is one fundamental principle to be observed in linking these different elements and drawing up the plan of the buildings: the strongrooms must be isolated from the rest of the building to ensure their protection against fire. This can be achieved either horizontally or vertically; put another way, the strongrooms can be built either physically separated from the other areas or separated from them by thick walls or placed above or below each other with fire resisting floors and ceilings.

The problems posed by such walls, floors and fireproof doors will be seen below. Here we need only consider the effects of this necessary isolation on the plan and general idea of the buildings. The most simple way to ensure the separation of strongrooms from other areas is of course to build the strongrooms block on one side and on the other the administrative and technical block, connected only by corridors with fire resisting doors.

However, this solution is not always practical if only because of the large surface area it requires. The isolation of strongrooms can be equally well achieved by means of fire resisting walls within the same block of buildings. It can also be achieved in vertical buildings by the use of fire resisting floors or ceilings.

Should strongrooms occupy upper or lower floors? It seems logical to place the heavier elements (strongrooms) in the lower parts of a budding; provided the lifts are adequate, working areas can then be on upper stories. This solution has the added advantage, in larger towns, of ensuring that offices, search rooms, etc. are furthest removed from street noises and polluted atmosphere. At street level or below would be areas for the reception and weeding of documents and certain workshops.

The other solution, which puts strongrooms above working areas, is technically feasible through the use of thick reinforced concrete floors supported by stout pillars. This allows working areas to remain at street level and avoids constant use of lifts.

The overall plan of the budding, then, depends on the shape and dimensions of the site and on the method adopted to separate the strongrooms from the other areas.

A traditional layout is L-shaped, with the strongroom block perpendicular or at right angles to the administrative and technical block. It is very satisfactory, as is the T-shaped layout which is similar.

Using fire resisting walls where necessary, concentric, square or rectangular layouts can also be adopted, the centre or one side or one angle of a quadrilateral being occupied by strongrooms.

If the budding consists of two blocks of strongrooms the most rational layout is either H-shaped or in the shape of a square U in which the median bar will represent the working areas.

If there are to be several strongroom blocks, a star shaped building will ensure the best operation of the service (plate 10b).

Provision for the Future: No archivist can claim to have done his job properly, when building or adapting a budding for his service, if he has not provided for extensions sufficient for the foreseeable future.

It is impossible to forecast the need for such extensions in the abstract; it obviously depends on the rate and extent of accessions, Careful statistics should be kept, to ascertain:

1. volume of transfers received during the decade preceding construction or adaptation;
2. size of transfers refused in recent years through lack of space;

3. yearly rate of population growth of the town or province (this rate, more or less, will have its repercussions on the number of files dealt with by offices);
4. approximate quantity of documents to be destroyed each year.

Ideally, an archive budding should, from the time of its construction or adaptation be designed to meet the needs of at least the next 10 years; in addition to the racking needed to store holdings at the time of building or adaptation, enough for the foreseeable transfers of the next 20 years should be installed, allowance being made for the corresponding destruction. But twenty years is a short time in the future of an archive service; one must think even further ahead.

Failing the initial construction of a building large enough to meet the needs of the next 50 to 100 years (and this would be excessively ambitious), provision for possible enlargement of both strongrooms and working areas should be made. This enlargement can be at ground level or vertically; in the latter case, provision should be made when the foundations are dug for future extra floors and weight to be placed on them.

In the absence of precise and detailed figures (a risky business anyway), it would be agreed, as a rule of thumb, that the potentiality of doubling the capacity of a repository would be looked upon as satisfactory.

The Open Plan and Flexibility

Flexibility is one of the qualities in the requirements of a library building which has high priority in the hierarchy of its disiderata. Library planning requirements and qualities have changed during this century, slowly at first but more rapidly recently with the changing needs of libraries and library users.

Although internal arrangements and library services vary from place to place, and from one type of library to another, recent libraries of all sizes have many common factors, which have been crystalised into the following ten desirable qualities-Faulkner-Brown's ten commandments. A library should be:

flexible	with a layout, structure and services which are easy to adapt;
compact	for ease of movement of readers, staff and books;
accessible	from the exterior into the building and from the entrance to all parts of the building, with an easy comprehensible plan needing minimum supplementary directions;
extendible	to permit future growth with minimum disruption;
varied	in its provision of book accommodation and of reader services to give wide freedom of choice;
organised	to impose appropriate confrontation between books and readers;
comfortable	to promote efficiency of use;
constant in environment	for the preservation of library materials;
secure	to control user behaviour and loss of books;
economic	to be built and maintained with minimum resources both in finance and staff.

These are the broad outlines of ten important qualities. Irrespective of size, these qualities can be applied in **varying** degrees. It is worth examining them in more detail.

Flexible: Flexibility of course does not mean that the structure is flexible and will bend or move under stress (that is reserved for the librarian). A flexible library building is one which permits flexibility in the layout of its planning arrangements, with structure, heating, ventilation and lighting arranged to facilitate adaptability. By arranging columns at regular spacing, or reducing the number of columns with long span beams, and by designing the floors to carry a superimposed live load of 150 lbs/ft^2 or 7.2 Kn/m^2 for bookshelf loading, it is easy to move departments, issue desks, bookshelves, reader places or other library functions to any part of the building.

Better flexibility is achieved when floors are level, without steps, and when the heating, ventilation and lighting allow rearrangement without the need for any alterations and yet maintain an adequate environment.

The planning arrangements can also be seen to be much more flexible if the number of walls in the building are reduced to a minimum, with the permanent walls concentrated in certain areas to form 'cores', containing the permanent features-stairs, lifts, toilets and ducts. Other walls, where security and privacy are absolutely essential, are not structural, and are designed to be demounted and erected elsewhere. The building and its components are designed to facilitate this. All other areas can be left open and by applying the well tried (but not universally accepted) experience of offices designed on 'burolandschaft' principles, visual and aural privacy are achieved very simply, with the bonus of much improved communications and supervision. The necessary visual privacy is achieved by varied furniture arrangements with bookshelves providing indigenous screening, and movable indoor planting additionally providing colour, a variety of forms and life to the interior.

Surprisingly, aural privacy is achieved by acoustic material on both the floors and ceiling, plus the introduction of an even level of ambient noise in the ventilation system. These factors ensure that noise levels of normal conversations are absorbed in a satisfactory manner, and are not distinguishable at distances of beyond 4 metres from source.

In an open-planned building designed flexible to cater to adaptations, relocation of departments and activities are achieved without having to resort to expensive contractual alterations, and the librarian is not inhibited from making changes or instituting experiments-they are achieved merely by moving furniture and bookshelves. If however the furniture is fixed or is built-in or built of brick, steel or reinforced concrete then it does present a more difficult problem. The furniture is immovable for all time, which assumes that needs will not change.

Furthermore it can be demonstrated that the open-plan flexible library can be economical in staff resources, since overseeing and informal control are faciliated by the openness rather than by dividing up the building into rooms or halls, thereby requiring less staff.

It has been demonstrated in many libraries that when shortage of recurring finance has resulted in severe inadequacies in staff numbers, whole departments of some libraries are unable to open because of lack of funds to employ staff (music departments are sometimes victims). In the case of open plan libraries, real economies in staff numbers can be made, without substantial reduction in library service. If the layout is planned so that staff are beneficially located to supervise and service more than one department, then the distressing conclusion to perhaps close, or in the case of a new library, not to open a particular department can be avoided. This will be assisted in an open plan library if, depending on size, the departments can be arranged contiguous with each other on the same level floor.

It can be seen therefore that the open plan has many advantages, that enclosed rooms disappear, or are drastically reduced in number, and that departments are in loosely defined areas, informally arranged in relationship to each other.

Compact: A compact building will assist the librarian in many ways. Theoretically travel distances will be reduced to a minimum if the building is a cube and on entry users are brought to the centre of gravity. Books, staff and readers will need to move shorter distances in a cubic building than in a linear building or one extended by moving away from a deep plan. There is also a bonus in economy of consumption of fuel and energy. This diagram compares deep and linear buildings of large floor area to demonstrate the difference in travel distance.

Accessible: The quality of lease of access' to the building and to the books is one to which much attention needs to be paid. An easy and inviting route to the entrance should also be unamibiguously defined. Once inside, the user should be aware of the location of the principal elements of the building-enquiries, the circulation desk, catalogues, tools for information retrieval, stairs, etc. and the routes should be strongly stated without an over-proliferation of signs and direction.

Extendible: Until recently all librarians and some architects have maintained that library buildings, especially academic libraries, are not finite. They should be capable of extension and land should be reserved for future expansion. A most significant development in British academic libraries is the recent report of a working party on Capital Provision for University Libraries-the

Atkinson Report. Among other things it recommends the adoption of the principle of a 'self renewing library of limited growth', and establishes new norms. This means that academic library buildings are to be finite with no provision for extension.

It is a commonly held view that every library building should be capable of extension, that the construction of the building will facilitate extension, and that at each stage of development the building should appear to be a complete entity. Naturally the choice of exterior materials and construction will be heavily influenced by this latter factor. The exterior wall of a library building can consist of a series of simple repetitive units which can be removed from the facade and re-used in an extended building. If the library is not extended it can stand in its present state as a finite, and apparently complete building. If the needs of the library change, the building can be changed reasonably easily. Some of the ten commandments can be bent, some diluted, but this one should not be abandoned.

Varied: The variety of book and of user accommodation in a library adds interest to the interior but also provides for the many needs and preferences of the users. This will vary considerably depending on size, function and location.

Organised: A principal quality or function of a library building is to organise the display of its library materials so that they are accessible and easily available. Simplicity in layout, arranged in an easily understood and inviting way is vital in both small and large libraries.

Comfortable: Before beginning the design of a library, the librarian and the architect together should visit a large number of libraries of all types. It is important to observe how libraries are actually used. Photographs and notes should diligently record this, and will probably include many cherished photographs of sleeping users. Almost without exception they will have occurred in large libraries with antiquated or inadequate ventilation without air-conditioning. A fresh, constant temperature and humidity not only promote efficiency of use, it encourages use. In some climates discomfort is caused if windows in a large library are opened-heat, cold, dirt and noise are offered 'open-access' from the external environment.

In other climates, to achieve the desirable comfort conditions it is important and economic to use the free facility nature offers from the external environment and induce it into the building with controls to regulate it according to need. Generally speaking this applies to large library buildings, especially those with a deep plan, and to those where study conditions can be offered with a secure aural environment. In smaller library buildings and in those without a critical acoustic need or a book preservation problem, the normal local custom of building can produce acceptable comfort conditions.

Nevertheless, in all libraries a good standard of lighting is necessary-there is a lot to be said for an evenly maintained level of about 400 lux at the working plane throughout the public areas. This will be adequate for most needs, including the illumination of the book title on the lowest shelf.

Constant in Environment: Reserach into the preservation of library materials indicates that a constant environment is necessary, and when this requirement is linked to the former-comfort of the users, an unvarying level of illumination, heating, cooling, ventilation and acoustics will give the type of environment needed in a library. The wall should be considered to be an environmental filter or regulator. It should reduce heat loss in winter and solar gain in summer. It should keep out intrusive external noises yet provide windows for prospect.

Secure: Security of the collections has always been of prime importance in libraries. The reduction of public access and egress to a single point well controlled by electronic or other means, the openness of planning to assist automatic overseeing of most areas goes some way to reduce the loss of books and to control the behaviour of users in many instances.

Economic: The energy crisis has hit all of us. Libraries can be expensive buildings to build and they can be expensive to run; in fact running costs have become a major financial consideration to librarians. In large libraries the deep compact plan requires long hours of artificial illumination and air-conditioning to create an even and constant environment. Every acceptable method must be examined to minimise cost without impairing service. In the first instance, when designing a building, economy in running costs can be effected by reducing the surface of the exterior skin of the building (walls and roof) as much as possible, so that the ratio of wall area to floor area is low. A building form with a cube shape is the ideal, but may not suit the library planning needs. However it is important that the building shape is as close to a cube as possible.

Secondly, windows allow heat to pass out of the building in winter and to pass into the building in summer from solar penetration. Window openings should be as small as possible and as a guide the recommended total area of window should not exceed 25% of the total wall area. Shaping the exterior of the building to provide shading for the windows can keep out solar penetration at the hottest part of the year, thereby reducing the cooling load in summer. There is no need to stress the importance of wall and roof thermal insulation.

Contrary to widely held belief, the great consumer of energy in a deep plan building in temperate climates is not the heating requirement in cold weather. Well insulated walls of minimum area are the only substantial source of heat loss. The centre part of the deep plan is not losing heat, since it is surrounded by a cocoon of warm air in the perimeter bay. In addition to the lighting the major consumers of energy are the fans to circulate air through the building and the refrigeration equipment to reduce the temperature in warm weather. The period when maximum energy is required is in hot weather with a full library, when the air-conditioning plant has to deal with high outside temperature, and with permanent artificial lighting to a high even standard.

If real economies in running cost are sought, then I suggest that a much closer look needs to be taken at the way in which our libraries are used. Some investigations on use carried out in three British University Libraries-Edinburgh, Newcastle and Nottingham-show the total weekly exit numbers recorded at the check-out points and give some indication of the intensity of use. The pre-examination revision period is highlighted in November and particularly in May. The most noteworthy factor is the understandable similarity of low use from June to October. This is the period when the air-conditioning has to work hardest to cope with high outside air temperature and solar gains on and through the building fabric. The first question to ask librarians to consider is-when briefing the architect it is possible roughly to quantify the anticipated average occupation of the building during the hottest part of the year. If the occupancy is low the size of the refrigeration plant in most libraries can be modest.

The second question is-in a deep plan building do all the lights throughout the building have to be left on during the summer, or can there be some logical means of switching arranged so that maximum economy can be effected?

The third is that if librarians can arrange most of the reading space near the perimeter might it be possible to depend on daylight, and therefore, in open plan flexible buildings, could seating be rearranged accordingly?

In conclusion, it can be seen that many of the advantages of an open plan library building, designed to give a high degree of flexibility, is a requirement of modern library planning and needs no justification. This strong requirement should be balanced with the other requirements and qualities of a library building.

It is hoped that economies in initial capital and recurring annual running costs resulting from open plan library buildings will be an additional beneficial factor.

10

Academic Library Materials Expenditures

INTRODUCTION

This article analyzes the impact of the rising cost of materials on academic libraries. While a number of papers discuss a particular library's difficulties or aspects of the problems faced by libraries, this article attempts to analyze the pressures and resources of academic libraries as a group, reflected in Association of Research Libraries (ARL) and Association of College and Research Libraries (ACRL).

Trends in total library expenditures and materials expenditures are compared with inflation indexes, the consumer price index (CPI), and the more specialized library price index (LPI). Inflationary trends and the impact of journal proliferation on library collections are discussed. If present trends continue, the cost of supporting a first rate library-one with on-site access to a wide range of current journals, a rich selection of recent monographs, and other reference and research materials especially in technical and scientific fields-will grow at a pace which is insupportable in the long run.

Data Sources

Much of the statistical data comes from one of four sources. ARL Statistics is an annual publication of the Association of Research Libraries, a federation of over 100 major libraries across the United States and Canada. ACRL University Library Statistics is an analogous publication of the Association of College and

Research Libraries which is published approximately every other year. In general, ACRL libraries are smaller than the ARL libraries but are similar in purpose. Price index data are extracted from Inflation Measures for Schools and Colleges: 1993 Update, an annual publication of Research Associates of Washington. Price data specific to the library world are from Library Journal's "Periodical Price Index."

Data Organization

The difference of scale between the typical university ARL library and ACRL library is such that they are best treated separately, although it will be seen that they are generally subject to the same trends and respond in similar ways. Connecticut and Brandeis are medium-sized members of the ARL and the ACRL, respectively. Connecticut, with 2.3 million volumes and a total budget of $13.1 million, swamps Brandeis, with 900,000 volumes and a budget of $3.8 million. In addition to university libraries, the ARL counts as members a dozen other research libraries. These libraries, listed in the end notes, are distinct from the university research libraries in that they are tasked to serve not just an academic circle but a community of national, if not international, scope. Unlike university research libraries, these are often directly funded by the federal government in the United States or Canada. Finally, there are also two nominally municipal libraries-Boston and New York Public-which transcend, with a combined total of 13 million volumes, the confines suggested by their geographic names. These libraries, whose activities are in some respects unique, are excluded from the analyses since the focus of this article is academic libraries.

Total Library Expenditures

The aggregate ARL library expenditures stood at $720 million by 1982, nominal expenditures more than doubled to over $1.5 billion in 1992. ARL library expenditures have increased between 5.69 percent (1991) and 10.26 percent (1985) per year during these ten years (except in the recession year of 1992, when expenditures climbed by only 3.65 percent). ACRL library expenditures seem to follow the same general trend, though at a lower level. The ACRL did not publish statistical abstracts for 1983, 1985, 1987,

1990, or 1992. No projection is made for 1992, and the expenditure values for the gap years are interpolated estimates. Still, it is evident that the nominal expenditures of the ACRL libraries have grown substantially, if not to quite the same extent as the ARL libraries. To allow for variation in the number of member libraries over the decade (especially in the ACRL, where fifteen new libraries reported total expenditure data in 1989), it is possible to look at total expenditures on a per library basis. On this basis, ARL expenditures have climbed from $7.12 million in 1982 to $14.1 million in 1992, a 98 percent increase in ten years. ACRL expenditures have increased from $2.63 million per library in 1982 to $3.52 million in 1989, a 34 percent increase in nine years.

The percentage growth in annual expenditures in the ARL and ACRL libraries with the consumer price index (CPI), the general measure of inflation, as well as a specialized library price index (LPI) for each year since 1982. The library price index is a measure of inflation as it affects libraries and is influenced by changes in pay scales for librarians as well as by changes in serial and monograph prices. The percentage increase in total expenditures for both the ARL and ACRL is generally greater than the CPI or LPI. Prior to 1987, ACRL expenditures generally kept even with inflation but grew faster than LPI in 1987, 1988, and 1989. Since 1989, however, expenditures have been in decline. On the whole, real expenditures in ARL and ACRL libraries have been increasing.

This is not to discount the fact that some libraries are falling behind, and the aggregate certainly hides the horror stories. At California Polytechnic State University, the library has dropped 330 current journals (of 3,313) over the last five years while increasing the journal expenditure by 82 percent to $700,000 per annum (Walch, 1993, p. 125). Staffing levels have been reduced by 24 percent over the last five years, from 70.5 FTE to 57 FTE, and monograph purchase has ground to a halt without a single new book being bought in three months. Roger Noll (1993) remarks that even Stanford, while able to purchase monographs, did not have the resources to catalogue and shelve the new books. Thus they sit in the basement, inaccessible and unused. But these are exceptionally unfavourable circumstances and do not reflect the norm. In ARL libraries, as the total expenditures have increased, so too has the proportion devoted to materials. In these libraries, the materials share has risen every year since 1982, steadily expanding from 31.10 percent of the total expenditures to 33.86 percent in 1992 (Association of Research Libraries, 1992). The steadiness of the increases in a population of over 100 libraries suggests a fundamental shift in group activity. Among ACRL libraries, such a shift of additional resources to materials is not apparent. This is, perhaps, because they are already so heavily committed to materials that there is no additional money for reallocation to materials. In absolute numbers, ARL libraries' material expenditures have climbed from $224 million in 1982 to $516 million in 1992, an increase of 130 percent. ACRL material expenditures are up by 84 percent. In contrast, the CPI has increased only 47 percent and the LPI by 72 percent during this period.

Serial and Monograph Expenditures

This expanded materials expenditure is increasingly devoted to serials. Figure taken from ARL Statistics, 1991-92, displays the increased relative weight given to serials. Funds allocated to purchase serials are growing swiftly and consistently. Since the graph is in constant 1982 dollars, a horizontal line indicates expenditures just keeping pace with inflation, and a downward slope indicates real decline. Serials expenditures started an explosive growth around 1986 and grew faster than total expenditures for nonserials materials. By 1992, serials show the greatest percentage increase of any category reported. Moreover, while total library expenditures and nonserial materials show signs of stabilization in the last year or two, serials expenditure continues to climb. Indications are that journal prices will continue to climb. Faxon and EBSCO have estimated 6.2 percent and 7.0 percent price increases, respectively, for journals in 1994. EBSCO had originally estimated a 10.5 percent price increase, but the unexpected strength of the dollar on the international market has mitigated a portion of that calculation ("Journal Prices to Rise..., 1993).

Figures compare the serial expenditures with that for monographs in both ARL and ACRL libraries. In contrast to the swift growth of the serial expenditures in these libraries is the nearly fixed level of expenditures on monographs. A fixed budget in inflationary times is, of course, declining in real terms. It may fairly be asserted that serials are crowding out monographs in library acquisition. Indeed, the share of material expenditures devoted by ARL libraries to monographs has fallen from 40.32 percent to 33.04 percent between 1986 and 1992. This seven point drop is a one-sixth decline in the portion of materials expenditures for monographs. Again, it is critical to recognize that even what seems to be a modest change in the percentage of budget can in fact carry major implications. This trend is as pronounced in the ACRL libraries, where monograph expenditures have fallen from 38.9 percent of the materials expenditures in 1986 to only 31 percent in 1991. These trends may be sustainable, simply altering the nature of library collections over time, were it not for the fact that even this tilt toward serials is not keeping pace with serial prices in certain disciplines. The problem, as discussed later, is that, as a category, serial prices are increasing rapidly, and libraries are being asked to subscribe to new journals.

The Race Between Expenditures and Prices

A way to examine the dynamics between rising expenditures on the one hand and the rising cost of materials on the other is to normalize both prices and expenditures to a common year and track the percentage change. Table displays indexes for select categories of library expenditures normalized to 1986 which serve as a baseline for subsequent analysis of prices. Indexes for specific categories of monographs such as U.S. college books, North American academic books, and median price of monographs purchased by ARL libraries, are presented in table; the indexes for serials of selected subject fields (for example, chemistry and physics, engineering, medicine), and for the general academic periodicals (United States and foreign) are presented in Table.

Monographs

Monographs cost more today than a few years ago, of course. The median price for monographs purchased by ARL libraries, up 44 percent in six years, has grown faster than either the CPI or LPI. U.S. college book and North American academic book indexes are almost exactly the same as the median price, as might be expected from the definitions (Research Associates of Washington, 1993) These three indexes have increased nearly at the same pace as the aggregate total library expenditures or aggregate materials expenditures of ARL libraries but have increased at a faster rate than monograph expenditures.

In spite of the growth in materials expenditure, spending on monographs does not seem to reflect higher monograph prices. As a result, many libraries are purchasing fewer monographs. ARL libraries have purchased fewer books each year from 1986 to 1990, recovering somewhat in 1991. During this period, these libraries were each buying 700 fewer books per year, an average decrease of 2.2 percent in the number of monographs purchased from year to year. The ACRL libraries were hit even harder for their size. While the ARL libraries managed to buy more monographs in 1991 and 1992, the average number of monographs purchased by ACRL libraries continued its downward trend, bouyed only in passing by the substantial expansion of the ACRL in 1989. Since 1989, both the median and total number of monographs purchased have fallen precipitously. The 1991 median (10,410) is 88 percent of 1989's (11,856). On the average, these libraries bought 180 fewer monographs each year since 1988.

As swiftly as monograph prices have risen, serial plus have risen even faster. In 1992, U.S. periodicals were 80 percent more expensive than in 1986. If serial expenditures are keeping pace with serial prices in general, it is because the price increases in some disciplines have been relatively modest. Annual price increases for fine arts journals have ranged from 4.0 percent to 5.5 percent per year for the last three years and are up 46 percent from 1986 to 1992 (row 4). During this period, history journals increased by 66 percent (row 5).

In contrast to the relatively modest price increases in humanities serials, stand the price increases in the scientific, technical, and medical disciplines (rows 6-8). The price increases have been greatest in chemistry and physics and in engineering. On the average, journals in 1992 were 108 percent more expensive than in 1986 in these fields and 82 percent more expensive in medicine. These fields drive serials expenditures. As shown in table, the CPI, in contrast, increased by a bare 27 percent over the same years; the LPI by 40 percent; the ARL libraries' total expenditures by 48 percent; material expenditures by 56 percent; and serial expenditures by 87 percent. The prices of scientific, engineering, and medical journals are racing ahead of all other measures of costs or expenditures.

IMPACT OF JOURNAL PROLIFERATION ON SERIAL EXPENDITURES

It must be the best of times and the worst of times in serials. Articles on serials cancellations abound. Indeed, fifty ARL libraries had plans to terminate over $7 million worth of serial subscriptions in 1992. Although there is a steady stream of reports of cancellations of current subscriptions, new serials emerge. Ulrich's reports that 3,800 new serials were launched since January 1, 1990 (Bowker International Serials Database, 1992, p. vii). The median number of current serials in ARL libraries has increased every year but one, climbing from 20,537 in 1985 to a peak of 22,287 in 1991 before declining to 21,750 in 1992. The aggregate number of serials held has climbed every year, from 2.9 million in 1985 to 3.1 million in 1992. The increased cost of providing adequate serial collections reflects both increased unit cost and a greater aggregate number of serials held.

Suppose, for example, that a library was committed to providing on-site access to all the titles listed in Index Medicus. The index listed 2,352 titles in 1992. To purchase most of these titles would have cost $660,722. This compares to a cost of $420,000 for 2,251 titles in 1988 which is an increase of 58 percent over five years. The average cost per title is up 53 percent, from $184.20 in 1988 to $280.92 in 1992 (Bowker International Serials Database, 1992, p. vii), the remaining 5 percent of extra costs being attributable to the larger number of serials. Similar circumstances surround serial collections in other fields.

The periodical price index surveyed seventy-six journals in mathematics in 1990. Just two years later the price index surveyed eighty-three titles, presumably as a reflection of the increased number of journal titles. The price index reported the average price of the journals had risen from $251.99 to $302.17 (Carpenter & Alexander, 1992, p. 57). But even more important than this 20 percent average price rise was the cost to maintain a full collection. To buy all seventy-six journals in 1990 would have cost $19,151. To buy all eighty-three in 1992 would have cost $25,080, an increase not of 20 percent but of just over 30 percent. Thus, not only are serials becoming more expensive, but there is tremendous pressure to add new serials to the collection.

Conclusion

ARL and ACRL libraries' data show that academic libraries have enjoyed an increasing level of support. Expenditures have increased in comparison to the broad measures of inflation-the CPI and LPI. The near doubling of ARL expenditures per library in ten years is evidence of a deep commitment to library resources. The critical issue for libraries and librarians, however, is that library expenditures have actually decreased with respect to the demands placed upon libraries by the constituencies they serve.

Decreasing real library expenditures on monographs deserves attention. There has been an absolute decline in the number of monographs purchased. Among the ARL libraries, monographs purchased have decreased from 3,006,538 in 1986 to 2,722,391 in 1990, while, among ACRL libraries, the number of monographs purchased has increased from 1,205,018 in 1986 to 1,251,531 in 1991. However, this increase in the ACRL aggregate total reflects an increase in the number of ACRL libraries not an increase in the average number of monographs purchased. Academic libraries are buying an ever smaller slice of the monograph pie.

Whereas the decline in the number of monographs purchased is absolute, there are more serials purchased from year to year. Still there is a decline in the proportion of serial universe covered. The repeated references to the cancellations of serials are evidence of the inadequacy of current funds to meet the needs of academic libraries. Although the literature abounds with reports of serial cancellations, the median number of current serials is actually up in ARL libraries from 20,537 in 1986 to 21,750 in 1992. Indeed, many of the patterns observed here have been evident since 1973. The ARL Serials Prices Project, published in 1989, reported:

> *The average funding support to ARL libraries rose 243% from 1973 to 1987 compared to a 182% rise in the U.S. CPI during*

the same period). Nonetheless, ARL libraries' average percent of expenditures devoted to materials rose from 29.2% to 33.1% with an accompanying shift in the percentage devoted to serials from 40.4% to 56.2%. During this same period, the average serials holdings of ARL libraries dropped from 32% of the estimated universe to 26.4%. According to data collected by ARL, in 1988 the median price of a purchased serial was $115.00-an increase of 32% since 1986. (ARL, 1989, p.2)

Since we excluded the twelve nonacademic ARL libraries, strictly speaking, a direct comparison between the ARL project and our study is inappropriate. However, over 90 percent of the libraries overlap. This said, since 1987, university ARL library expenditures are up 37.7 percent while the CPI is up only 24.3 percent-a rate of real increase in expenditures from 1987 to 1992 of 2.2 percent per year. Expenditures of the ACRL libraries have increased as well. The steady upward creep of the share of total expenditures devoted to materials, evident in all ARL libraries since 1973, has continued. Within the material expenditure category, the proportion devoted to serials has continuously grown. This unprecedented level of spending on serials within ARL libraries is only four points behind that of the ACRL libraries, which have historically devoted a greater share of their material budget to serials.

It is clear that libraries are under stress generated by the longterm increase in the number and cost of serials. Library expenditures, although growing in real terms, are not keeping pace with the demands upon them. The libraries seem to have adapted to the situation so far, if at the cost of holding a reduced slice of the pie, both in serials and in monographs.

THE WHOLE COST OF LIBRARIES-LIBRARY FINANCE

In addition to the operating budget, which may also include endowment and similar special funds, there may also be capital budgets. The interaction among these budgets is seldom stated, though many academic institutions have begun the practice of including some kind of operating endowment in fund drives meant to finance new construction. Most institutional budgets are aggregated from budget requests submitted by various agencies who may or may not have engaged in any prior consultations. The fragmentary nature of institutional budgeting makes it clear that the true cost of running a library is not simply that shown in the operating budget.

Different kinds of libraries receive different kinds of budgetary treatment. In general, public library budgets are the most comprehensive, since they operate autonomously and have to be responsible, for example, for utility and cleaning costs, which seldom show up in an academic library budget. Special libraries sometimes do not have a direct budget, and their costs are paid from various sources as appropriate or charged back to various accounts. Some academic budgets make provision for staff benefits, while in other cases these are charged against a central fund. These variations exist whatever budget style is used-from the simplest line item to the most complicated programme budget. This makes consideration of total library costs a very complex matter, but, in view of the need for exercising the highest level of budgetary restraint, it is essential to know what these costs are.

Building and Maintenance Costs

Among the most important costs that do not usually show up directly in an academic or special library budget are those related to the building and its equipment. These include utility costs-heat, light, and power-which may be included in another part of the institutional budget, somewhat in the manner of overhead. Given the size of most libraries and the fact that they are open long hours with sizable populations, it is clear that building maintenance costs will also be sizable. It is, therefore, strange that these costs are seldom taken into consideration when planning new or renovated libraries.

Added power consumption or changes in heating, ventilation, and air conditioning (HVAC) requirements are seldom considered when planning for the installation of automated systems, despite the fact that concentrations of machines and people tend to generate both heat and noise beyond that expected in the simpler days when most libraries were planned. This is particularly important when the need is to fit systems into an existing building. Most library budgets include lines for door guards, and some include at least a part-time position responsible for building security, but few budgets (other than public libraries) include the cost of personal security or of cleaning, though both can be substantial in a large library. In part, this is because these activities are the responsibilities of other parts of the organization, which are themselves differently organized. It may be possible to extract costs associated with the library, but this kind of effort tends to be more expensive than the results are worth. It also results from the fact that different reporting mechanisms result in the activities being under the charge of different senior officers, so that there may be little or no interaction in the course of budget decision making.

Personal security is usually the responsibility of another agency, though there has been a move in some institutions to involve a wider range of administrators. Whereas the actual costs of repairs to the building may be charged back to the library budget, it is not often seen as parallel that the costs of personnel safety incidents should also be charged back. In fact, because of the nature of the building's use and the value of the materials housed, most libraries do make de facto assignments concerned with personal safety without showing the cost separately. Again, because they operate separately, large public libraries and museums have arrived at much better estimates of such costs and include them in their budgets. Because such considerations can affect the ways in which libraries (or individual departments such as Special Collections) can operate, their incorporation into the library budget, or at least consultation about these costs, would make clearer the actual cost of operation and enable better decisions to be made about how to do business.

Most library budgets will carry some line within the budget covering the cost of repairs even if the amount is small since there will inevitably be power and equipment failures and broken furniture. In this, library budgets differ from other parts of an academic institution, primarily because, whereas classrooms are shared by many departments, these are single users and the costs of replacement and repair can be allocated directly. The amount so allocated tends to have historic roots and acts as a kind of amortization fund on the theory that it is likely that a certain number of chairs and tables will wear out each year and can be replaced individually rather than wholesale which tends to be the method used for classroom buildings.

The increasing amount of electronic equipment in libraries has begun to strain the operating budget since very few institutions create sinking funds to cover the inevitable cost of replacement with more up-to-date equipment. Testimony to this is the increasing number of idle computer terminals in many libraries awaiting replacement or repair, also the long delay in upgrading computer systems beyond the time their peak usefulness has passed. Although it has long been the practice to provide a budget line for equipment service contracts (such as typewriter cleaning or maintenance for microform readers), the increase in the number and kinds of equipment has far exceeded the capacity of this usually modest part of the budget. Even the relatively simple need to keep terminals and workstations clean has been overlooked as a cost although it is as important as programming or CPU maintenance. In fact, libraries are finding that the cost of going electronic is far more complex than was ever thought.

Insurance for library buildings and collections can be very expensive, and insurance companies are now insisting on better building conditions, including the provision of proper disaster control systems. This may prompt institutions to think more thoroughly about one of their most expensive and valuable investments. Renovation of a library can be extremely costly, but the replacement of lost collections can be prohibitive, and may, in some cases, be impossible. There are no simple ways to keep the valuation of a library's collections up to date.

It is possible to use annual expenditures to increase total worth, but there are few algorithms to upgrade the cost of the replacement of older materials, and the calculation of the cost of replacing bibliographic data, whether in paper or electronic format, defies any existing system, since the original costs are either lost in time or composed of so many separate operations that the calculation of a total cost may prove impossible. Although they are referring principally to the concept of depreciation, both Christianson (1992) and Carpenter and Milican (1991) stress the importance of including processing costs. Only when a disaster wipes out a library do most institutions realize the cost of replacement which must cover not only the direct cost of materials but also the cost of processing them-a cost that is not covered by any insurance policy.

In the light of these facts, it is clear that more should be spent on preventive measures, including preservation, but such a change runs counter to most academic or other library budget styles. In the same way as repairs to a building are likely to be delayed until the cost demands either a separate fund drive or a legislative appropriation, the repair of the collection may well be delayed until there is virtually no collection to repair. Many libraries maintain minimal budgets for replacement, but, for the bulk of the collection where the deterioration is slow and silent, most libraries must seek special grants or look to national programmes which use microforming or digitization. Overlooking such needs leads to the undercapitalization of the library.

Building and maintenance costs can thus be seen to cover a wide range of costs, some of which are recognized, some of which are not. Again some are included, even if inadequately, in the library budget; some are the responsibility of other agencies and may or may not be provided for in their budgets; others are not covered at all. These costs include:

* repairs, whether major or minor;
* maintenance contracts for equipment;
* utilities;
* cleaning;
* insurance;
* amortization funding for new equipment;
* collection maintenance and preservation;
* personal safety costs; and
* disaster preparedness costs,

Growth Costs

It is in the nature of libraries to grow. Predicting the rate of growth is not an easy task (Drake, 1976). There have been several attempts to control growth in favour of stable library size. These efforts are generally more successful in smaller libraries where the principal need is to support teaching rather than research (Gore, 1976). In larger libraries, the need for specific items in the collection may diminish over time. The famous Pittsburgh study (Kent et al., 1979) simply confirmed this but did not explain how to predict what items will be used. In many subject areas, books and periodicals simply do not go out of date; in others, use may diminish to the degree that they do not need to be retained. However, even in technologicals fields, new uses have been found for older materials-for example, in exploring its history or in recovering an earlier base on which to measure change as in ecological studies. While it is clear that comprehensive libraries are not needed at all locations, there are no clear paradigms by which to determine what to keep and what to discard. Use studies have a role to play and can certainly help direct future growth. However, use also reflects the fashion of the day. Only too of often a researcher, seeking to probe new fields, finds that the materials needed have been discarded. Popular culture students often find themselves in this situation (Brooks, 1993) or those who want to revisit older times and interests (Heinzkill, 1990; Metz & Foltin, 1990).

Administrators and librarians who have pinned their hopes (for cutting budgets and keeping the library building smaller) on electronics and cooperative schemes have usually had those hopes dashed by the complexities of each alternative. In any event, both depend for success on the original items having been kept somewhere. It is unlikely that many libraries can be assured that anything they discard will automatically be available to them through interlibrary loan or document delivery. There is no current evidence that libraries are coordinating serial cancellations (Martin, 1992; Price & Carey, 1993), which suggests that interlibrary cooperation is still a hit or miss affair. There are also transfer costs associated with heavy reliance on document delivery. Leach and Tribble (1993) suggest that libraries will begin to invest more of their budgets in delivery services rather than purchase, but they also raise many issues related to financing and managing this very different kind of library operation (pp. 360-64).

In such circumstances, libraries can be expected to grow, whether by the addition of printed materials or by the incorporation of electronic alternatives (which still require space), and administrations will continue to find that they need new buildings or at least better ones. One solution is the construction of storage libraries which are cheaper to construct than regular libraries and can also make more effective use of environmental controls than buildings which are heavily used daily. There are, of course, processing and retrieval costs associated with running a separate building, but these are less than similar costs associated with an increasingly larger library building. Another alternative is the use of compact shelving, whether within the existing library or as an adjunct structure. User-accessible compact storage works best with smaller frequently used collections but can also be used in remote storage facilities where access is under staff control. Compact shelving offers the opportunity to house more materials in less space but also carries new operating costs, somewhat akin to the older system of paging that was used with closed access library stacks. There are also some questions about the speed with which materials can be retrieved, and such forms of storage raise questions related to equal access by handicapped persons. Here the cost benefits of storage have to be weighed against any resulting user service costs.

Because libraries usually occupy prime space, any expansion is likely to be costly even if it is underground. This has made the idea of an electronic library very attractive since it appears to offer the chance to house more in less space. In fact, the changeover to electronic access may well need every bit as much space as the traditional expansion of the printed collections, since the user space will have to grow proportionately to use. One architect has claimed, for instance, that workstations for computer-related work will need between fifty and sixty square feet of space, as against the twenty to twenty-five square feet that is now provided for a reading space (Jeffrey Freeman, personal communication, 1990). A major research library reference area must now include many more reader spaces of a larger size than was ever intended in the original design. Because electronic information does not yet include all publishing, the traditional collections are unlikely to diminish to make room for them. What this means is that the basic design concepts behind library buildings are changing rapidly, and older buildings cannot easily adapt to the new needs.

Traffic patterns will change and tend to concentrate the user populace more. Many libraries have also found that the electronic media have led to a new need for instruction space, separate from regular user space, since library instruction is no longer simply a matter of being shown where reference works and other tools, such as the catalogue, are located. The combined effects result in a library very different from that to which most administrators have been accustomed, and it is difficult to persuade them that the budgets that support them must also change. If libraries are to provide adequate services to users, they have to maintain adequate staff ratios, and these tend to remain constant. Some of these issues were examined as long ago as 1969 (Knight & Nourse, 1969) and later by Baumol and Marcus (1973), and the various budget ratios and patterns have not changed substantially since that time. Economies of scale are not readily available to a library where the transactions remain individual and unique. This remains true even in the use of electronic information. Such complexities make the modern library a much more difficult building problem, one which is barely now beginning to be addressed by librarians, administrators, and architects. The costs associated with changing building needs are seldom conveniently placed within existing budget paradigms, if only because it is difficult to place them clearly within categories.

* Are CD-ROM workstations capital equipment or the equivalent of periodical subscriptions?
* Have they replaced some elements of the older traditional budget or simply added new ones?
* How does one calculate the added utility costs of new electronic equipment, and where should they show up in the budget?
* Are there other support costs that must be included in the budget?
* What are the costs of different space alternatives?
* What alternative uses could be made of any money saved on building construction?
* How far can mechanical and electronic retrieval systems replace staff costs, or will they simply be added budget items?

These are only a few of the budgetary questions associated with the "new" library. Formerly, most such costs showed up in the "other" category of support expenditures, traditionally, about 10 percent of the total, but this is changing rapidly (Hayes, 1982; Kantor, 1986; Budd, 1990). Many libraries are now spending more than this proportion on electronic systems alone without taking into consideration more mundane daily expenditures. Without a substantial infusion of new money, an increase of that order can only be attained at the expense of other budget items-notably library materials-and there is a limit as to how far such a process can go before the library becomes dysfunctional.

Electronics

Too many have seen the advent of long-distance electronic information transfer not only as a way of extending the services a library can provide but as a way of cutting costs. As several speakers at the Computers in Libraries Conference (Oakland, California, 1991) pointed out, such an attitude overlooks the very real cost of telecommunication, the costs of staff training, and the substantial costs for equipment and installation. Many of these issues are discussed in Campus Strategies for Libraries and Electronic Information (Arms, 1990) but with little attention to budgetary effects. Since most wide area and local area networks are handled on an institution-wide basis, these are seldom charged back to individual operational units. This may change as the Internet and similar networks are privatized, leading to direct user charges. The internal result may well be similar to the change that was made in telephone billing when central overhead costs were charged back to individual units based on their share of the total system. This is likely to come as a shock to most users since networks have been thought of as essentially free. Institutions, on the other hand, which have tried to update their communications-for instance, by laying fibre optic cables-have come to realize that there are large capital costs and ongoing maintenance costs. Usage costs, in the form now familiar for telephones, have not yet emerged clearly but are certain to be developed either in an attempt to control usage or to recover costs.

Cost Recovery: Libraries have already had to grapple with this kind of issue in the provision of online services. Discussion of cost recovery has largely been conducted under the rubric of "Fee or Free," though, as White (1993) has pointed out, this is a misleading approach since, in fact, everything has a cost and has to be paid for. It is only a question of who will pay, and where the money will come from. Similar reservations were raised by Nielsen (1989) who was concerned at the relationships being drawn between cost and values. Taylor (1984) presented a very convincing case for fees for database searches using the analogy of photocopy services. The latter costs were, for a while, provided free until libraries realized: (1) that the cost would swamp the budget and (2) that photocopy provided a good additional income source. Whether the service is provided internally or by contract, it is now customary for there to be a user charge. Taylor predicts that database searching and other analogous electronic services will also require charges if only to regulate use and prevent a drain on the budget. White, as cited earlier, cautions that there are problems in trying to distinguish between traditional and new services-the moral basis on which librarians justify charges-but he does not deny that the services cost money.

The problem is compounded as libraries move toward including such services in their online catalogue systems. Many system vendors now make a great point of ways in which their systems link to other databases and services, such as the UnCover document delivery service. The advent of direct user searches and the possibility of using credit cards for payment is tending to eliminate such activities from those of the library proper.

It is therefore becoming difficult to draw the lines among library, departmental, and personal budget expenditures. Even if all such services became payment driven, someone would still have to provide the space and the equipment, tend to the hardware and software involved, and provide instruction when needed. Should these services be charged for, and, if so, who would pay? Should there be an overhead for each transaction or should the parent institution provide these through a central budget? These issues are still unresolved. Libraries seem to have engaged in ad hoc planning and to have drawn money from wherever possible. Nor has the issue of handling income from fees and charges been resolved, though libraries seem to be encouraged to charge for more and more services. Warner (1990) offers some suggestions for resolving such issues, though these relate more to special libraries.

Cost Centres and Overhead: As programme or functional budgets have become more accepted, libraries have begun to look at the concept of cost centres. The new electronic services can well be so regarded, with the caveat that these are linked to other more traditional services, such as reference and circulation, because of their side effects on those operations. Defining library cost centres is difficult, except in the case of standalone operations like interlibrary loan, while it is possible to argue that technical services as a whole is a kind of overhead. This introduces a new aspect of overhead costing, which has not been customary other than in special libraries.

There are sizable overheads in any library. These include general administration, supplies, systems support (from the library and the institution), and (in such cases as online services or bibliographic instruction) part-time assignments of staff together with benefits and support. To these can be added any direct system or vendor charges-*e.g.*, for maintenance or upgrading. The result is a budget considerably different from a line item budget or even a simple programme budget. If indeed all overhead or associated costs-such as heat, light, and power-and general administration were added, it would also be considerably larger than the traditional programme budget. Despite the growth of such costs in any institution or library, there has been a move (mostly from federal programmes) to lower the definition of overhead so that costs associated with grant projects may no longer be adequately recovered.

This has had an indirect effect on libraries, which had been seen as part of that overhead in that they have to continue to provide the necessary services from diminished budgets since the parent institution is no longer receiving the same reimbursement. It is true that many libraries were never allocated research overhead directly and may never have received the amount they used to justify, but this does not vitiate the argument that, in the new electronic era, libraries must be much more concerned with indirect and overhead costs.

User-Related Costs: Although it has never been the custom to count user costs as part of the library budget these are a real cost to the parent institution which must pay for the time used by its employees. If a considerable part of that time is used in walking to and from the library with no apparent return (the book wanted is out), then that time is wasted. Here electronic systems can play a part in developing higher returns on user time. Online circulation information, particularly when accessible through office computers, can help users plan library visits more fruitfully.

This information also makes it possible for users to ask for materials to be held at We circulation desk, thus reducing everyone's expenditure of time Dahlgren (1990) outlines many of the elements that should be considered when choosing a circulations system including user costs and benefits. It is also possible to load reserve book lists and thus to update these online quite apart from being able to give information about actual usage, which can help in determining retention on the list thus making the whole operation much more cost effective from both the library and the faculty point of view. Online catalogue searching can also be linked to interlibrary loan or to document delivery. In this way, online information can play a significant role in streamlining both library and user activity.

The budgetary effects of this improvement are diffused and unlikely to show up directly in a budget line, but indirectly these can help to refine collection management and reduce lost user time. This topic is mentioned here to encourage libraries and administrators to look beyond the actual budget figures when making decisions. The examination is akin to a user environmental impact study and has some of the same difficulties-notably converting such savings into dollar figures. But the attempt can and should be made since automation is usually presented as saving money without any concrete evidence (Martin, 1986). If user time were seen as a library cost element, then savings in that time would be seen as actual rather than illusory savings. Leaving the user out of the budgetary calculation is rather like a business ignoring customer preferences.

All library activities should be re-examined from a user point of view. For the most part, these activities are designed with the library staff in mind, which may be fine internally but overlooks whether these best serve the user. This may or may not cost the library more-double staffing for both reference and information desks, for example-but it will result in better use, which is in the best interests of both the library and its parent institution. It may also result in a realignment of some expenditures-*e.g.*, the transfer of some staff members from internal circulation to document delivery or an increase in levels of staff when it is realized that the circulation desk handles a regular quota of reference questions. It may, on the other hand, be possible to close a service station altogether as a reflection of use patterns. Even so simple a matter as closer attention to signage (usually a minimal budget item) can result in better usage patterns and a better use of the budget available.

Value Maintenance: Financial accounting systems for colleges and universities and for public sector organizations in general are constructed under the rules of generally accepted accounting principles, as shown in the various guidelines composed by the National Association of College and University Business Officers and similar organizations. Most library studies, excellent though they are, on economic theory (Schauer, 1986), on accounting methods (Smith, 1991), or on budgeting practice (Trumpeter & Rounds, 1985), are written without taking explicit account of the institutional context. The assumption seems to be that this is a given, whereas, in fact, it can have a substantial impact on what the library can or cannot do.

The principal aim of these accounting systems is to record accurately what the assets and liabilities of the organization were at the beginning of a period, what they were at the end, and what activities occurred between those points in time to cause the changes. In their attempt to be entirely factual, these accounting systems focus on actual rather than projected or estimated values. Thus they record the value of assets such as buildings only at the original price paid. Any subsequent expenditures for enlargement or restoration are simply added to the original recorded value regardless of any changes which may have occurred in the value of those dollars. Similarly, they make no attempt to recognize that assets may grow in value over time, nor that replacement costs may be significantly different in summary, generally accepted accounting standards make no provision for recognizing the current value of an asset to the institution.

Similarly, operating budgets are solely concerned with current expenditures, and capital budgets are developed to take care of necessary current expenditures-*e.g.*, repairs or expansions. All these financial concepts are important tools for fiscal control, but these need to be supplemented by "management accounting" concepts. Of these, the most important concept is value maintenance.

In principle, the idea is fairly simple. Librarians or other administrators want to maintain the current value of library assets to the ongoing life of the institution or constituency. There are two dimensions to this concept: (1) maintaining the current value of the assets (buildings, collections, etc.); and (2) since institutional needs evolve over time, modifying those assets over time so as to maintain their usefulness. These two dimensions may be thought of as upkeep and renewal.

There are three classes of asset with which librarians are concerned: facilities, collections (or, more broadly, access to information), and equipment.

Buildings deteriorate over time, as a function both of use and of decay. Each building can be thought of as a series of "systems," such as the foundations and walls; roof and windows; electrical, plumbing, HVAC; floor and wall coverings; and so on. Each system has a con and a life cycle. For example, the roof on a library may cost $100,000. Depending on the materials used and the climate, it may be necessary to replace it every twenty-five to forty years. Based on the cost and the life cycle, it is possible to estimate what amount should be put aside each year so as to be able to replace it when needed. The sum of the amounts needed for each system is the total amount that should be budgeted each year for asset upkeep. It is estimated that such a provision should be in the range of 1 to 1.5 percent each year.

The second dimension of value maintenance recognizes the effect of change. Alteration in the mix of users or changes in the methods of pedagogy or in technology can result in demand for more or less user space, for different kinds of space, or for additions to space. In addition, libraries have a special problem in dealing with growing collections. The "renewal" component of value maintenance can be very substantial and may require budgeting 1 or 2 percent per year of the replacement cost of the facility. Together these dimensions imply setting aside as much as 4 percent annually of the replacement cost-a very substantial addition to the usual operating budget.

The same concepts can be applied to library collections. Although all institutions recognize that their collections are extremely valuable, only recently have some institutions begun to assign an asset value to their collections. In part, this attitude has resulted from the fact that library materials purchases are made from current operating budgets and not seen as a capital expenditure. Whether or not the collections are recorded as a capital asset, it is essential to maintain their current value.

Upkeep is the primary concern. As with a building, the total collection can be thought of as a series of collections, each with different costs and life cycles. This is most clear in the sciences where the currency of the information is critical. Such collections have a very short life cycle, needing to be "replaced" yearly, and the retention of older materials adds a significant housing cost. Other collections, such as literature and language, do not deteriorate as quickly. These collections do need to be refreshed by adding current publications, but the whole collection remains useful and may even grow in value over time. From an analysis of the needs of each collection, the "upkeep" portion of the value maintenance budget may be calculated. This calculation can be used as a factor in budget construction and allocation.

Unkeep, however, is not enough. New programmes, changes in curricula, or the development of new reader interests require "renewal" expenditures. These expenditures are major and easy to overlook when planning new programmes and research projects.

The rapid growth of electronic access to information adds complexity to the problem. Such access comes at a cost, which has been regarded as an added operating expense. From a "management accounting" perspective it may be more useful to view it as part of the cost of maintaining the current value of the library as an information asset. Finally, the concept of value maintenance can be applied to library equipment-increasingly electronic equipment. The life cycles of the equipment are so short and the new technologies expanding so fast that the distinction between upkeep and renewal is less significant though still useful. Since the life cycles are only from three to five years, it is vital that library budgets make annual expenditure or reserve provisions to enable regular and frequent replacement of equipment.

The basic point is that institutional budgeting and accounting systems make it more difficult, rather than easier, to understand and provide for the whole cost of libraries. Librarians and administrators need to understand the management accounting approach of value maintenance, and to budget on that basis. By allowing for the upkeep and renewal of facilities, collections, and equipment, we can come closer to fulfilling our responsibilities. Nothing in this approach, of course, makes any new funding available, but it does make it easier to demonstrate the need for additional resources and helps in the better allocation of the available resources.

Without pretending to have engaged in an exhaustive analysis here, the goal has been to show that there are many unconsidered costs in running a library. Unless these are considered, changes and improvements may not have the desired effects. Many organizational decisions are made without a clear understanding of the financial effects, some of which may be delayed and others of which may be external to the library. The result can be a less than successful library programme.

During the winter of 1990, the author was invited to join Babson College as its first Chief Information Officer (CIO). Concurrently, the entire college community was in the throes of a detailed self-evaluation culminating in a new strategic plan. This process helped to restate and clarify Babson's mission as an educational institution dedicated to the development of innovative leaders capable of initiating, managing, and implementing change. Furthermore, the college committed itself to teaching with a global perspective And to the integration of information technology into all aspects of the Babson learning experience (Babson College, 1991).

To achieve the latter objective, the chief information officer was charged with creating and maintaining a "real world" information resource environment for the use of students, faculty, and administrators. In establishing this new office, the Babson College brought together the entire campus's existing information service departments, including academic computing, administrative computing, media services, the Babson College Telephone Company (BABTELCO), and the library. These functional areas had never operated in concert before. Each reported to a different senior administrator; possessed its own personnel structure, policies, and procedures; and provided services according to its own sense of customer requirements. To fashion a new environment within which to realize the Babson's strategic objectives, the CIO was obliged to reshape the operating units that now reported to him, provide his staff with a common sense of mission, and instill in them a sense of customer service that transcended their specific job assignments.

The organization that emerged from this effort was named the Information Technology and Services Division (ITSD). Its newly defined mission and strategic plan, which emerged from an intense discussion process involving both ITSD personnel and its customers, began as follows:

Consistent with the overall strategic plans of the College, it is the mission of Babson's Information Technology and Services Division (ITSD) to provide in partnership with the Babson Community information and services to proactively support the educational programmes, operation requirements, and business plans of the College. To achieve these ends through innovation and excellence, ITSD will deploy the best in proven information technologies.

In brief, the ITSD intended to deliver on this challenging assignment through the innovative use of an integrated information services organization. The model for such a structure, the so-called "Information Utility," was already present in private industry and was in fact emerging in leading U.S. colleges and universities. Babson's information resource management members are adapting this approach to their own institutional settings and in so doing are providing their colleagues with a practical illustration of how to effectively restructure information services to enhance performance and competitive advantage.

This article examines the forces at work within the modern organization that are driving information professionals to reconsider how best to structure and deliver their services. Global information needs, the increasing diversity and complexity of available information resources and systems, and the escalating "utility" costs of service maintenance are all factors influencing these developments. The author therefore begins with a consideration of external environmental forces and the emergence of the "knowledge worker" as the IRM professional's primary customer. From this more general discussion, the author will focus upon the positioning of the library within the context of the information utility model. The conclusion will Provide readers with some thoughts on the critical success factors associated with integrating the library into the I/U.

A Select Vocabulary

Though from a sister discipline, the author views the challlenges of IRM through a different lens than that of the typical library administrator. His use of terminology may not always appear, therefore, to be appropriate (or recognizable) to his audience. To orient the reader for the discussion that follows, and position the frame of reference away from the established library science framework of concepts and responsibilities and more toward a comprehensive information resource management perspective, the following terms and definitions are offered.

1. end-user-Also referred to as "customer," "patron," or "constituent," the end-user is the knowledge worker in the modern organization. I/T systems, services, and resources must be tailored to the requirements of the end-user who in turn addresses through his/her efforts the Primary mission of the parent organization.
2. enterprise-While "enterprise" may be used interchangeably with organization" and "institution," it is the preferred term because it conveys action and the creation/delivery of value to the end-user. Regardless of the strategic focus of the organization, enterprises must create "value" as

perceived by their customers if they are to survive and prosper. This statement applies to government services and higher education as well as private industry. Similarly, the "library" must be viewed as an enterprise within the "information utility" which is itself an enterprise within the parent organization.

3. information resource management-the economical and efficient management, servicing, and support of all information (in whatever format) that is of value to the organization. The value-added component of IRM is the information utility's ability to deliver accurate specific information to the end-user in a timely manner.
4. IRM strategic planning-IRM strategic planning is a necessary subset of the parent institution's process. It is necessarily shaped by the goals and objectives of the greater organization and must complement the more global directives established in the corporate plan.
5. information services Professional-While the terms librarian, archivist, records manager, and systems analyst have relevance in today's information technology environment, the twenty-first century information utility requires the services of cross-trained, highly integrated staffs of I/T professionals to act as facilitators, catalysts for change, standards monitors, and resource managers for complex user-driven and controlled information delivery systems.
6. information utility-Within any organization, the information utility includes all of those resources, services, and facilities that comprise, process, and deliver information to the end-user. More than computer hardware and software, an information utility is an approach to customer service that emphasizes availability, ease of access, economy, efficiency, and accountability to the community.
7. knowledge-We often think of "information" in terms of documents, records, files, etc., but these are merely formalized vehicles for the delivery of data to an end-user. Historically, these information products were/are self-supporting and generally sufficient in terms of satisfying

the needs of the end-user. With recent developments in I/T, "information" alone is not satisfactory-primarily because there is too much of it and the "products" in question are insufficiently focused and unadaptable. Instead, users seek "knowledge": a higher level of information, at times in multimedia formats, tailored and processed to address a specific requirement. Correspondingly, knowledge tools, such, as artificial intelligence systems and hypertext databases, facilitate the manipulation of information to meet end-user needs.

8. knowledge worker-This is the end-user who employs a wide range of information technologies to draw upon diverse information resources in a variety of formats to address his/her immediate needs through the sophisticated researching, sifting, search, and reassembly of data into highly usable formats. Note that all "knowledge workers" are "end-users" but not all "end-users" are "knowledge workers." It is the responsibility of information service professionals to assist in the development of end-users into knowledge workers.

9. strategic planning-Strategic planning is that process of thought and action that directs the long-term growth of an organization. It focuses upon the clearly defined mission, goals, and objectives of the organization; assesses the available resources to bring these milestones to fruition; and establishes a method of performance measurement. The rigor of the process places considerable demands on management but is essential to corporate prosperity and hence to the interest of all stockholders (*i.e.*, organization members and those served by the organization).

By way of orientation, This exhibit graphically represents the flow of raw data in various media and formats to intermediate data collection and distribution platforms (*e.g.*, databases). From there, it is manipulated by higher-level information processing ("knowledge") tools (*i.e.*, computer applications) and then transmitted via an array of networks to the desk top of the end-user, who, in this illustration, is either a living person or an automated process. The ultimate delivery of "knowledge" as defined earlier may then lead to specific informed actions. As described here, the entire set of transactions in constitutes modes of information resource management and use within the modern organization. The unique character of these processes defines the institutional context and corporate culture within which people work. Like the role of IRM itself, the modern organization is also changing radically due to technological innovation and adaptation. To better understand the forces at work and what they mean to the library administrator, we will next explore the information requirements and evolving I/T environment of the modern institution.

Organizational Environments and IRM

As we proceed toward the twenty-first century, organizations are becoming less bureaucratic, more complex, and global in their orientation. Their management structures will flatten with senior executives playing a larger role in the direct management of people and processes. These players will map out the strategic programmes for their organization, employing external alliances, resource sharing, outsourcing (*i.e.*, the use of external agencies to perform services or processes hitherto maintained by the organization), and new information technologies to enhance their overall performance. Middle management will grow thin and serve primarily as a group of technical specialists developing policies, procedures, and applications for other employees. The vast majority of those remaining will directly contribute to value creation in terms of either products of services provided to the customers of the organization. In this more fluid, less hierarchical environment, most, if not all, employees will have both information resource management and production responsibilities.

Information technologies have played, and will continue to play, a central role in this restructuring of the enterprise. They facilitate streamlining and encourage a more entrepreneurial operating mode among managers now freed from dependence on others for vital information. For example, through electronic mail and executive information systems, senior managers can readily access field personnel and assess the status of far-flung projects. The management process need not occur through direct face-to-face interaction but may be mediated through electronic mail and teleconferencing. These same technologies also tend to foster linkages with external global partners. As operations become more complex, they are being segmented with the relocation of specific functions to the most advantageous locales. For example, automobile and computer manufacturing now occurs in a global arena where plants are located near cheap labour and the necessary raw or processed materials. Without the computer and telecommunication facilities of the modern corporate infrastructure, these arrangements would not have materialized.

Furthermore, the ubiquitous and increasingly user-friendly nature of emerging information technologies has meant that line managers rather than technologists have taken charge of the resource, refocusing IRM requirements on core services and strategic business objectives (Tom, 1991; see also Emery, 1987; Targowski, 1990). This trend exemplifies the realization that, to manage a process, those in charge must also control the related IRM functions. It manifests itself in the growing acceptance of end-user "ownership" of the data and even associated information systems and IRM resources. The proliferation of I/T and information resources throughout the organization is illustrated in Figure. No functional area in this representation of an organization is without its IRM capabilities and responsibilities.

Each operating unit of the XYZ Organization has fully integrated business functions. This structure is indicative of the worker empowerment and managerial flattening of the enterprise alluded to earlier. Similarly, each unit has its own information processing; capabilities, ranging from individual personal computer workstations to large corporate databases run on mainframe computers. They also have access to, if not complete control over, the I/T tools, hard copy and online information resources, and associated support services deemed necessary to satisfy the requirements of their customers. The organization's administrative units are similarly endowed as the "owners" of human resource, financial, real estate, purchasing, insurance, and other corporate data. In this context, the information services arm of the organization acts as the I/T standards watch dog, the keeper of networks and operating environments (*i.e.*, the I/T infrastructure), the provider of Access to external information utilities (*e.g.*, bibliographic utilities and extracorporate electronic mail networks), and the developer and supporter of new I/T capabilities.

As a result of these functional allocations of I/T responsibilities, information service Providers within the modern organization are concerned less with the efficient and economical storage of data and more with the proactive delivery of knowledge. Thus the IRM shopping list includes such products as intelligent, personal computerbased tools for end-users; future-focused decision support systems; business simulation software; and expert systems (for an excellent summary of what leading international CIOs are looking for, see The Index Group, 1991. See also The Index Group, 1988). Throughout, the objective of these I/T scenarios is to empower the end-user and to put this person in touch with the appropriate data to address immediate customer needs today and plan for tomorrow.

Bibliography

Appleby, Paul H.: *Policy and Administration*, Birmingham, AL: University of Alabama Press, 1949.

Augustine, P.A.: *Social Equality in Indian Society*, New Delhi, Concept, 1991

Austin, Granville: *The Indian Constitution: Cornerstone of a Nation*, Oxford, Clarendon Press, 1966.

Barbour, Ian G.: *Issues in Science and Religion*, New York, Harper & Row, 1966.

Bhatia, B.M.: *Indian Agriculture: A Policy Perspective*, New Delhi, Sage, 1988.

Burch, William R. Jr.: *Human Ecology: An Environmental Approach*, Belmont, Duxbury Press, 1994.

Chandel, M.P.S.: *A Social Force in Politics: Study of Scheduled Castes of UP*, New Delhi, Mittal, 1990.

Charles Howard McIlwain: *Constitutionalism: Ancient and Modern*, N.Y., Cornell University Press, 1958.

Crick, F.: *Life Itself: Its Origin and Nature*, New York, W.W. Norton, 1982.

Dalrymple, G. Brent: *The Age of the Earth*, Stanford, Stanford University Press, 1994.

Daubenmire, R. F.: *Plants and Environment*, New York, Wiley, 1947.

Doolittle, R. F.: *Probability and the Origin of Life, in Godfrey*, New York, W.W. Norton & Co., 1983.

Ehrlich, Paul R: *Human Natures: Genes, Cultures, and the Human Prospect*, Washington, Island Press, 2000.

Francis, E.: *Bureaucratic Power in National Politics*, Boston, Little Brown and Company, 1978.

Gay, Kathlyn: *Saving the Environment: Debating the Costs*, New York, Franklin Watts, 1996.

Gentry, Robert V.: *Creation's Tiny Mystery*, Knoxville, Earth Science Associates, 1986.

Kruvant, W.J.: *People, Energy, and Pollution*, Cambridge, Ballinger Publishing, 1975.

Michael, B.: *The Politics of Succession in India*, Connecticut, Greenwood, 1976.

Palumbi, Stephen R.: *The Evolution Explosion: How Humans Cause Rapid Evolutionary Change*, New York, Norton, 2001.

Paul, H.: *Policy and Administration*, Birmingham, AL: University of Alabama Press, 1949.

Prahlad, K.: *Governance and Public Administration for Poverty Reduction*, Salvador, Brazil, 1997.

Price, C. C.: *Synthesis of Life*, Stroudsburg, Hutchinson & Ross, 1974.

Sharman, R.: *Aspects of Political Ideas and Institutions in Ancient India* Delhi, Motilal Banarsidass, 1968.

Unwin, David M.: *The Pterosaurs from Deep Time*, N.Y., Pi Press, 2005.

Vallette, J.: *The International Trade in Wastes: A Greenpeace Inventory*, Washington, Greenpeace, 1989.

Vegotsky, A., and Fox, S. W.: *Protein Molecules: Intraspecific and Interspecific Variations*, New York, Academic Press, 1962.

INDEX

♦♦♦